Anatomy

1600 Multiple Choice Questions

Anatomy

1600 Multiple Choice Questions

M. J. T. FITZGERALD, M.D., Ph.D., D.Sc.
Professor and Chairman, Department of Anatomy, University College, Galway, Ireland; Formerly Lecturer in Anatomy, St. Thomas's Hospital Medical School; and Associate Professor, Department of Biological Structure, University of Washington School of Medicine, Seattle

JAMES P. GOLDEN, M.B., F.R.C.S.
Lecturer, Department of Anatomy, University College, Galway, Ireland; Formerly Resident in Surgery, St. Joseph Hospital, La Branch, Houston, Texas

MAEVE FITZGERALD, M.B., B.Ch., B.Sc.
Lecturer in Histology, University College, Galway, Ireland; and Formerly at the Department of Oral Biology, University of Washington School of Dentistry and Department of Anatomy, St. Louis University School of Medicine, St. Louis, Missouri

Butterworths
London and Boston

THE BUTTERWORTH GROUP

ENGLAND
Butterworth & Co (Publishers) Ltd
London: 88 Kingsway, WC2B 6AB

AUSTRALIA
Butterworths Pty Ltd
Sydney: 586 Pacific Highway, NSW 2067
Melbourne: 343 Little Collins Street, 3000
Brisbane: 240 Queen Street, 4000

CANADA
Butterworth & Co (Canada) Ltd
Toronto: 2265 Midland Avenue,
    Scarborough, Ontario, M1P 4S1

NEW ZEALAND
Butterworths of New Zealand Ltd
Wellington: 26–28 Waring Taylor Street, 1

SOUTH AFRICA
Butterworths & Co (South Africa) (Pty) Ltd
Durban: 152–154 Gale Street

USA
Butterworth
161 Ash Street, Reading, Mass. 01867

All rights reserved. No part of this publication may be reproduced or transmitted in any form or by any means, including photocopying or recording, without the written permission of the copyright holder, application for which should be addressed to the publisher. Such written permission must also be obtained before any part of this publication is stored in a retrieval system of any nature.

This book is sold subject to the Standard Conditions of Sale of Net Books and may not be re-sold in the UK below the net price given by Butterworths in their current price list.

First published 1973
Second Impression 1975

© The Authors 1973

ISBN 0 407 50395 1

Suggested U.D.C. Number: 611(076)

Printed in England by Compton Printing Ltd.,
Aylesbury, Bucks.

CONTENTS

Preface vii
Instructions for Students ix
Section   I   Upper Limb (Questions 1–169) 1
Section   II   Lower Limb (Questions 1–192) 23
Section   III Thorax (Questions 1–165) 47
Section   IV   Abdomen (Questions 1–214) 65
Section   V   Pelvis and Perineum (Questions 1–131) 85
Section   VI   Head and Neck (Questions 1–325) 101
Section   VII Nervous System (Questions 1–266) 135
Section   VIII General Histology (Questions 1–120) 171
Section   IX   General Embryology (Questions 1–41) 185

PREFACE

The increasing use of multiple choice examinations at both undergraduate and postgraduate levels has stimulated the production of this book.

Each of the authors has postgraduate clinical experience, and clinical relevance has been the criterion for selection of the questions in gross anatomy and neurology. The topics in histology have been chosen with an eye to their bearing on important concepts in physiology and general pathology. We judge embryology to be fundamental to the understanding of gross anatomy; and its relevance to obstetrical and paediatric practice is also quite clear.

Gross anatomy, embryology and histology are merely separate aspects of body structure, and we have combined all three as much as possible in order to preserve this integrity.

Six types of question have been selected. Four of these are classical multiple choice, in which appropriate selections are made by the student from groups of answers provided. This form is traditional in the United States. The fifth type is the group true—false, which is widely used in Great Britain. It has the advantage (from the examiner's point of view!) that a correct answer cannot be reached indirectly, by a process of elimination.

Finally, some illustrations have been included to add variety to the format.

M.J.T.F.
J.P.G.
M.F.

INSTRUCTIONS FOR STUDENTS

The various sections of the book should be used upon completion of the corresponding parts of your Anatomy course.

In assessing your own performance on a particular section you should bear in mind that the questions fall into two categories for scoring purposes. In the 'group true—false' type of question you are required to judge whether each one of the five statements provided is true or false. In grading these answers it is common practice to award a mark for each correct decision, and to deduct a mark for each wrong decision. The average score obtained by chance alone will therefore be zero.

The remaining questions belong to the second category, in which you are required to make a selection from a number of answers provided. A score of 20—25 per cent is to be expected on the basis of chance.

SECTION 1 — UPPER LIMB

QUESTIONS 1-35:

FOR EACH OF THE FOLLOWING MULTIPLE CHOICE QUESTIONS SELECT THE *ONE* MOST APPROPRIATE ANSWER:

1. THE SPINE OF THE SCAPULA IS CONTINUED LATERALLY AS THE:
    A. Coracoid process
    B. Angle of the scapula
    C. Infraglenoid tubercle
    D. Supraglenoid tubercle
    E. Acromion process

2. ACTION OF PECTORALIS MAJOR:
    A. Flexion of humerus
    B. Medial rotation of humerus
    C. Adduction of humerus
    D. All of the above
    E. A and B only

3. THE VEIN WHICH PIERCES CLAVIPECTORAL FASCIA:
    A. Basilic
    B. Lateral pectoral
    C. Internal thoracic
    D. Axillary
    E. Cephalic

4. THE AXILLARY VEIN:
   A. Is lateral to the axillary artery
   B. Is devoid of valves
   C. Lies anterior to pectoralis minor
   D. Is directly continuous with the brachiocephalic vein
   E. None of the above

5. THE THORACODORSAL NERVE:
   A. Is a branch of the posterior cord of the brachial plexus
   B. Supplies the serratus anterior muscle
   C. Is cutaneous to dorsal surface of thorax
   D. All of the above
   E. A and B only

6. THE SUPRASCAPULAR NERVE:
   A. Supplies the infraspinatus muscle
   B. Arises from the lateral cord of the brachial plexus
   C. Notches the axillary border of the scapula
   D. All of the above
   E. A and B only

7. THE HUMERUS MAY BE ROTATED LATERALLY BY:
   A. Subscapularis
   B. Supraspinatus
   C. Pectoralis major
   D. Deltoid
   E. None of the above

8. THE MUSCLE PAIR RESPONSIBLE FOR ABDUCTING THE HUMERUS TO A RIGHT ANGLE:
   A. Deltoid and subscapularis
   B. Deltoid and supraspinatus
   C. Supraspinatus and subscapularis
   D. Teres major and subscapularis
   E. Deltoid and teres major

9. THE MUSCLE PAIR WHICH ASSISTS IN ELEVATING THE ARM ABOVE THE HEAD:
   A. Trapezius and pectoralis minor
   B. Levator scapulae and serratus anterior
   C. Rhomboid major and serratus anterior
   D. Rhomboid major and levator scapulae
   E. Trapezius and serratus anterior

10. THE SHOULDER JOINT IS WEAKEST:
    A. Above
    B. Below
    C. In front
    D. Behind
    E. Laterally

11. NERVE(S) SUPPLYING SHOULDER JOINT:
    A. Radial
    B. Musculocutaneous
    C. Axillary
    D. Suprascapular
    E. All of the above

12. THE NERVE TRUNK MOST INTIMATELY RELATED TO THE CAPSULE OF THE SHOULDER JOINT IS:
    A. Radial
    B. Axillary
    C. Median
    D. Ulnar
    E. Musculocutaneous

13. THE FOLLOWING MUSCLES BELONG TO THE 'ROTATOR CUFF' GROUP *EXCEPT:*
    A. Subscapularis
    B. Deltoid
    C. Supraspinatus
    D. Infraspinatus
    E. Teres minor

14. MUSCLE HAVING AN INTRACAPSULAR TENDON:
    A. Long head of biceps
    B. Short head of biceps
    C. Coracobrachialis
    D. Long head of triceps
    E. None of the above

15. MUSCLE INSERTED INTO ANTERIOR ASPECT OF HUMERUS:
    A. Subscapularis
    B. Supraspinatus
    C. Infraspinatus
    D. Teres minor
    E. None of the above

16. MUSCLE INSERTED INTO MEDIAL LIP OF INTERTUBERCULAR SULCUS:
    A. Teres major
    B. Teres minor
    C. Pectoralis major
    D. Pectoralis minor
    E. Latissimus dorsi

17. THE APEX OF THE CUBITAL FOSSA IS FORMED BY:
    A. Brachioradialis and pronator teres
    B. Brachialis and pronator teres
    C. Brachioradialis and biceps brachii
    D. Biceps brachii and supinator
    E. Brachioradialis and supinator

18. THE LATERAL CUTANEOUS NERVE OF THE FOREARM IS DERIVED FROM THE ---------- NERVE:
    A. Musculocutaneous
    B. Median
    C. Ulnar
    D. Radial
    E. Axillary

19. MUSCLE(S) INNERVATED BY THE ANTERIOR INTEROSSEOUS NERVE:
    A. Flexor pollicis longus
    B. Pronator quadratus
    C. Flexor carpi radialis
    D. All of the above
    E. A and B only

20. FLEXION OF WRIST JOINT IS PERFORMED BY:
    A. Flexor carpi radialis
    B. Flexor digitorum superficialis
    C. Flexor digitorum profundus
    D. All of the above
    E. A and B only

21. TENDON DIRECTLY MEDIAL TO DORSAL (LISTER'S) TUBERCLE OF RADIUS:
    A. Extensor pollicis brevis
    B. Extensor pollicis longus
    C. Extensor indicis
    D. Extensor carpi radialis longus
    E. Extensor carpi radialis brevis

22. TENDONS PASSING SUPERFICIAL TO THE RADIAL ARTERY:
    A. Biceps and brachioradialis
    B. Brachioradialis and flexor carpi radialis
    C. Abductor pollicis longus and brachioradialis
    D. Abductor pollicis longus and extensor pollicis brevis
    E. Extensor indicis and extensor pollicis longus

23. THE CARPAL BONES ARTICULATING WITH THE RADIUS ARE:
    A. Scaphoid and pisiform
    B. Lunate and pisiform
    C. Lunate and trapezium
    D. Lunate and scaphoid
    E. Scaphoid and capitate

24. THE TRIANGULAR FIBROCARTILAGE:
    A. Is attached to styloid process of radius
    B. Separates synovial cavities of radiocarpal and inferior radio-ulnar joint
    C. Articulates with lunate bone when wrist is adducted
    D. Is stationary during pronation and supination
    E. Is commonly absent

25. METACARPAL BONE WITH EPIPHYSIS AT PROXIMAL END:
    A. First
    B. Second
    C. Third
    D. Fourth
    E. Fifth

26. THE FLEXOR RETINACULUM IS ATTACHED TO:
    A. Proximal carpal bones
    B. Distal carpal bones
    C. Radius and ulna
    D. All of the above
    E. A and B only

27. THE CARPAL TUNNEL CONTAINS:
    A. Flexor carpi ulnaris tendon
    B. The ulnar artery
    C. The radial artery
    D. Palmaris longus tendon
    E. None of the above

28. ABDUCTION OF THE THUMB CARRIES IT:
    A. Forwards away from the palm
    B. Backwards to the side of the palm
    C. Towards the index finger
    D. Laterally, away from the index finger
    E. In a direction intermediate between A and D

29. NUMBER OF MUSCLES INSERTED ON INDEX FINGER:
    A. Three
    B. Four
    C. Five
    D. Six
    E. Seven

30. DORSAL INTEROSSEI DIFFER FROM VENTRAL INTEROSSEI IN HAVING:
    A. A different manner of origin
    B. A different manner of insertion
    C. A different nerve supply
    D. No action on interphalangeal joints
    E. Flexor action on the metacarpo-phalangeal joints

31. THE INTRINSIC MUSCLES OF THE HAND ARE ALL SUPPLIED BY:
    A. The median and ulnar nerves
    B. The first thoracic segment of the spinal cord
    C. The lateral and medial cords of the brachial plexus
    D. All of the above
    E. A and B only

32. IN THE HAND, THE MEDIAN NERVE SUPPLIES:
    A. Abductor pollicis brevis
    B. Adductor pollicis
    C. First dorsal interosseous
    D. Abductor pollicis longus
    E. Extensor indicis

33. THE INNERVATION OF THE LUMBRICAL MUSCLES IS RELATED TO THE INNERVATION OF:
    A. Flexor digitorum superficialis
    B. Flexor digitorum profundus
    C. Extensor digitorum
    D. The interossei
    E. The two flexor carpi muscles

34. THE SKIN OF THE INDEX FINGER IS SUPPLIED BY:
    A. Ulnar and radial nerves
    B. Radial and median nerves
    C. Median and ulnar nerves
    D. Median only
    E. Radial only

35. THE SKIN OF THE PALM IS SUPPLIED BY:
    A. Ulnar and median nerves
    B. Radial and median nerves
    C. Radial and ulnar nerves
    D. Ulnar nerve alone
    E. Radial nerve alone

QUESTIONS 36–53:

THE SET OF LETTERED HEADINGS BELOW IS FOLLOWED BY A LIST OF NUMBERED WORDS OR PHRASES. FOR EACH NUMBERED WORD OR PHRASE SELECT THE CORRECT ANSWER UNDER:

   A. If the item is associated with A only
   B. If the item is associated with B only
   C. If the item is associated with both A and B
   D. If the item is associated with neither A nor B

   A. Supinator of forearm
   B. Flexor of elbow
   C. Both
   D. Neither

36. Biceps brachii
37. Triceps brachii
38. Pronator teres
39. Brachialis
40. Coracobrachialis

A. Biceps
B. Brachialis
C. Both
D. Neither

41. Origin from humerus
42. Insertion into radius
43. Blood supply from brachial artery
44. Motor supply from median nerve

A. Anterior interosseous nerve
B. Posterior interosseous nerve
C. Both
D. Neither

45. Pierces supinator muscle
46. Derived from ulnar nerve
47. Sensory to wrist joint
48. Extensive cutaneous distribution

A. Flexor of wrist
B. Flexor of index finger
C. Both
D. Neither

49. Flexor carpi radialis
50. Flexor carpi ulnaris
51. Flexor digitorum superficialis
52. Flexor digitorum profundus
53. Palmaris longus

QUESTIONS 54–69:

DIRECTIONS: In the following series of questions, one or more of the four items is/are correct. Answer A if 1, 2, 3 are correct; B if 1 and 3 are correct; C if 2 and 4 are correct; D if only 4 is correct; and E if all four are correct.

54. MUSCLES ARISING FROM THE CLAVICLE INCLUDE:
    1. Pectoralis major
    2. Trapezius
    3. Deltoid
    4. Subclavius

55. DIVISION OF THE LONG THORACIC NERVE IS MANIFESTED BY:
    1. Inability to retract the scapula
    2. Wasting of the pectoralis major muscle
    3. Weakness of humeral adduction
    4. 'Winging' of the scapula

56. MUSCLE(S) ATTACHED TO MEDIAL EPICONDYLE OF HUMERUS:
    1. Flexor carpi ulnaris
    2. Pronator quadratus
    3. Palmaris longus
    4. Flexor pollicis longus

57. EXTENSOR(S) OF ELBOW JOINT:
    1. Anconeus
    2. Brachioradialis
    3. Triceps
    4. Extensor carpi ulnaris

58. THE FLEXOR CARPI RADIALIS MUSCLE:
    1. Is a flexor of the wrist
    2. Is an abductor of the wrist
    3. Is supplied by the median nerve
    4. Grooves the trapezoid bone

59. BRACHIORADIALIS MUSCLE:
    1. Arises from lower half of humerus
    2. Inserts into distal end of radius
    3. Is a flexor of elbow joint
    4. Is supplied by the median nerve

60. THE POSTERIOR INTEROSSEOUS NERVE:
    1. Passes between the radius and ulna
    2. Lies on the interosseous membrane throughout its course
    3. Is cutaneous to the back of the hand
    4. Supplies the extensor digitorum muscle

61. THE ULNAR NERVE SUPPLIES:
    1. The medial half of flexor digitorum superficialis
    2. The lumbrical to the little finger
    3. The abductor pollicis brevis
    4. The first dorsal interosseous muscle

62. JOINT(S) CONTAINING INTRA-ARTICULAR FIBRO-CARTILAGE:
    1. Sternoclavicular
    2. Acromioclavicular
    3. Wrist
    4. First carpometacarpal joint

63. ARTICULATING SURFACE(S) IN WRIST JOINT:
    1. Proximal surface of pisiform
    2. Head of ulna
    3. Distal surface of scaphoid
    4. Distal surface of radius

64. MUSCLE(S) SUPPLIED BY ANTERIOR INTEROSSEOUS NERVE:
    1. Flexor digitorum profundus
    2. Flexor pollicis longus
    3. Pronator quadratus
    4. Pronator teres

65. DIGITAL SYNOVIAL SHEATH(S) GENERALLY CONTINUOUS WITH ULNAR BURSA:
   1. Index finger
   2. Middle finger
   3. Ring finger
   4. Little finger

66. GIVE(S) ARTERIAL CONTRIBUTION TO DEEP PALMAR ARCH:
   1. Main radial artery
   2. Main ulnar artery
   3. Deep branch of ulnar artery
   4. Superficial branch of radial artery

67. MUSCLES INNERVATED BY THE MEDIAN NERVE INCLUDE:
   1. Palmaris brevis
   2. Opponens pollicis
   3. Adductor pollicis
   4. First lumbrical

68. EXTENSION OF THUMB IS AIDED BY:
   1. First lumbrical
   2. First palmar interosseous
   3. First dorsal interosseous
   4. Abductor pollicis longus

69. INTEROSSEI INSERTED INTO MIDDLE FINGER:
   1. Second palmar
   2. Second dorsal
   3. Third palmar
   4. Third dorsal

QUESTIONS 70–94:

THE GROUP OF QUESTIONS BELOW CONSISTS OF NUMBERED HEADINGS, FOLLOWED BY A LIST OF LETTERED WORDS OR PHRASES. FOR EACH HEADING SELECT THE *ONE* WORD OR PHRASE WHICH IS MOST CLOSELY RELATED TO IT.
NOTE: EACH CHOICE MAY BE USED *ONLY ONCE*.

70.	Coracobrachialis	A.	Flexion of humerus
71.	Supraspinatus	B.	Lateral rotation of humerus
72.	Subscapularis	C.	Abduction of humerus
73.	Pectoralis minor	D.	Medial rotation of humerus
74.	Infraspinatus	E.	None of the above

75.	Common interosseous artery	A.	Subclavian
76.	Profunda brachii artery	B.	Axillary
77.	Suprascapular artery	C.	Brachial
78.	Radialis indicis artery	D.	Ulnar
79.	Lateral thoracic artery	E.	None of the above

80.	Trapezius	A.	Musculocutaneous nerve
81.	Deltoid	B.	Accessory nerve
82.	Brachialis	C.	Ulnar nerve
83.	Supinator	D.	Axillary nerve
84.	Flexor carpi ulnaris	E.	None of the above

85.	Flexion of the elbow	A.	Median nerve
86.	Abduction of the shoulder	B.	Musculocutaneous nerve
87.	Extension of the elbow	C.	Radial nerve
88.	Pronation of the forearm	D.	Ulnar nerve
89.	Abduction of index finger	E.	Axillary nerve

90.	Flexion of wrist	A.	Extensor digitorum and extensor digiti minimi
91.	Extension of wrist		
92.	Extension of fingers	B.	Flexor carpi radialis and flexor carpi ulnaris
93.	Adduction of wrist		
94.	Abduction of wrist	C.	Flexor carpi ulnaris and extensor carpi ulnaris
		D.	Flexor carpi radialis and extensor carpi radialis longus
		E.	Extensor carpi radialis and extensor carpi ulnaris

QUESTIONS 95-149:

IN REPLY TO THE FOLLOWING QUESTIONS INDICATE WHETHER YOU THINK EACH STATEMENT IS *TRUE* OR *FALSE*:

THE SERRATUS ANTERIOR:
95. Arises by digitations from the lower eight ribs
96. Inserts into the axillary border of scapula
97. Acts synergistically with trapezius in abduction of arm to 90°
98. Nerve supply from thoracodorsal nerve
99. Paralysis gives rise to the condition known as 'winged scapula'

THE DELTOID:
100. Initiates abduction of the shoulder
101. Has an extensive range of action because it is multipennate
102. Is supplied by the radial nerve
103. Inserts into a rough elevation on lateral aspect of humerus
104. Together with the head of humerus, it is responsible for the characteristic roundness of the shoulder

THE AXILLARY ARTERY:
105. Begins at the upper border of the clavicle
106. Terminates as it crosses the inferior border of pectoralis minor
107. Is contained in the axillary sheath
108. Has the median nerve anterior to its proximal third
109. Has the radial nerve behind its distal third

AT THE SHOULDER JOINT:
110. Bony surfaces permit considerable movement
111. Stability depends mainly on glenoidal labrum
112. Subscapularis bursa communicates with synovial cavity
113. Long head of triceps arises within joint
114. Head of humerus is entirely intracapsular

THE HUMERO-ULNAR JOINT:
115. Is a synovial joint
116. Resembles an interphalangeal joint in being a hinge joint
117. Shares joint cavity with superior radio-ulnar joint
118. Capsule is stronger anteriorly and posteriorly than at the sides
119. Triceps is attached to capsule

THE BICEPS BRACHII:
120. Flexes both the shoulder and the elbow joints
121. Both supinates and pronates the forearm
122. Short head arises from clavicle
123. Tendon of long head is partially enclosed in synovial membrane
124. Is supplied by the median nerve

THE RADIAL NERVE:
125. Arises from lateral cord of the brachial plexus
126. Supplies brachioradialis
127. Divides near the elbow into muscular and cutaneous components
128. Innervates most of the dorsal skin of the hand
129. When injured, gives rise to the condition known as 'wrist drop'

THE ULNAR NERVE:
130. Arises from the medial cord of the brachial plexus
131. Supplies skin on medial side of arm and forearm
132. Passes behind medial epicondyle of humerus
133. Supplies first dorsal interosseous muscle
134. When injured, thenar muscles are wasted

THE MEDIAN NERVE:
135. Arises directly from trunks of the brachial plexus
136. Crosses the axillary artery from lateral to medial side
137. Enters the forearm through pronator quadratus
138. Enters the hand by passing through carpal tunnel
139. When injured gives rise to the condition known as 'claw hand'

THE CARPAL TUNNEL:
140. Is a fibro-osseous tunnel formed by carpal bones and palmar aponeurosis
141. Contains the tendons of flexor digitorum superficialis
142. Contains both the radial and ulnar arteries
143. Compression of nerve in tunnel causes sensory loss in index finger (palmar surface)
144. Contains portion of ulnar bursa

THE PALMAR APONEUROSIS:
145. Is attached to the skin of the palm by fibrous septa
146. Is attached distally to the fibrous flexor sheaths
147. Protects the underlying tendons
148. Receives tendon of palmaris longus
149. Apex is attached to flexor retinaculum

POSTERIOR ASPECT OF SHOULDER REGION

IDENTIFY THE NUMBERED STRUCTURES:

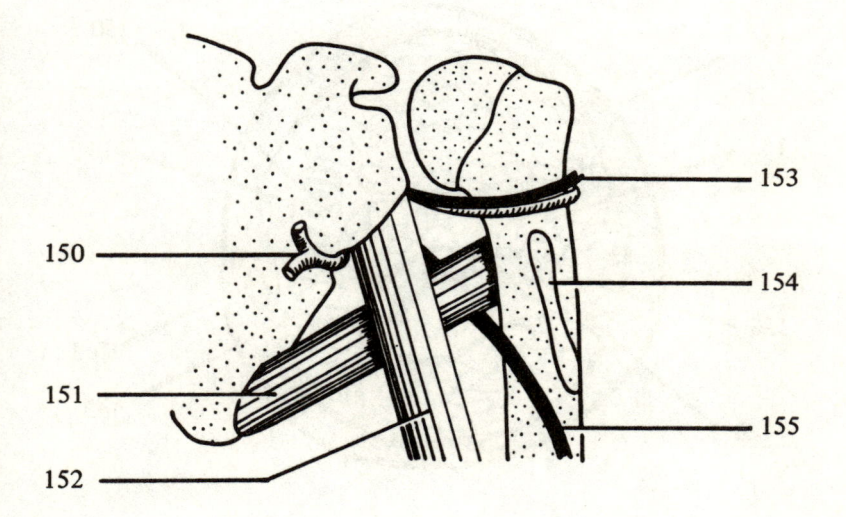

A.	Triceps, long head	G.	Rhomboid major
B.	Ulnar nerve	H.	Circumflex scapular artery
C.	Biceps, long head	I.	Triceps, lateral head
D.	Axillary artery	J.	Teres minor
E.	Triceps, medial head	K.	Radial nerve
F.	Axillary nerve	L.	Teres major

17

SECTION THROUGH LEFT ARM

IDENTIFY THE NUMBERED STRUCTURES:

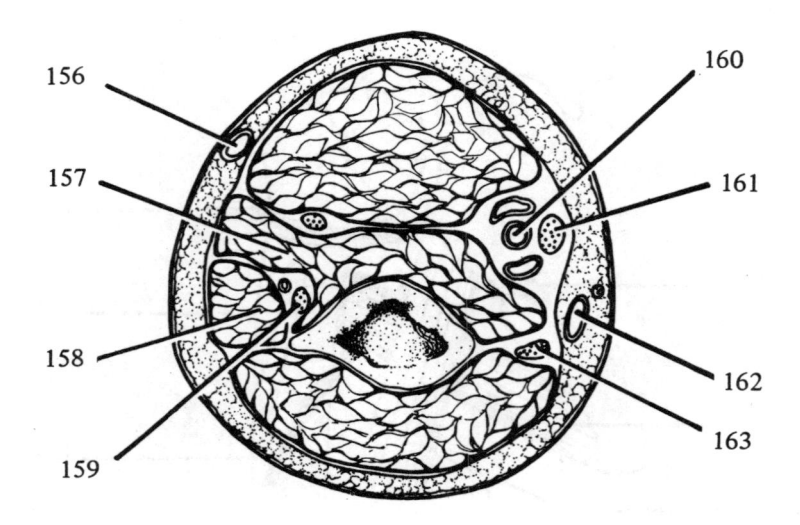

A. Biceps brachii
B. Cephalic vein
C. Radial nerve
D. Brachialis
E. Posterior interosseous nerve
F. Brachial artery
G. Medial cutaneous nerve of forearm
H. Ulnar collateral artery
I. Triceps
J. Vena comitans
K. Ulnar nerve
L. Brachioradialis
M. Basilic vein
N. Median nerve
O. Profunda brachii artery
P. Musculocutaneous nerve

18

TENDONS AT WRIST

IDENTIFY THE NUMBERED STRUCTURES:

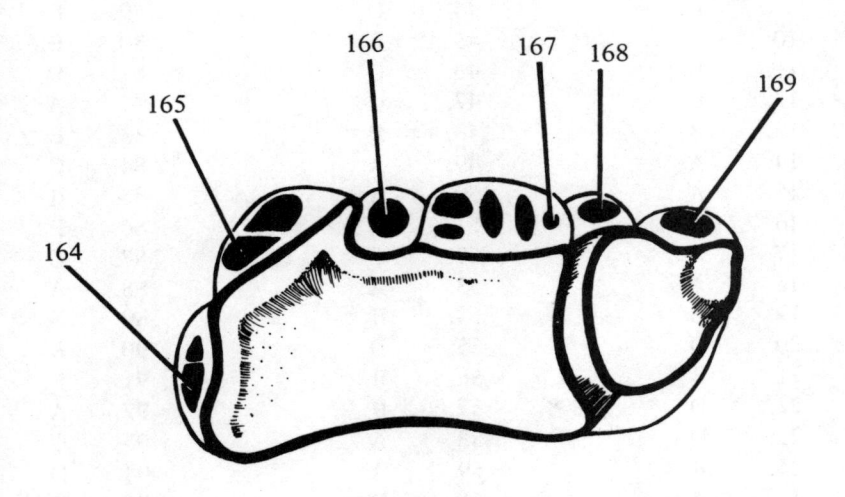

A. Flexor carpi radialis
B. Abductor pollicis brevis
C. Extensor indicis
D. Extensor carpi radialis brevis
E. Extensor carpi radialis longus
F. Brachioradialis
G. Abductor pollicis longus
H. Extensor carpi ulnaris
I. Extensor pollicis longus
J. Extensor pollicis brevis
K. Extensor digiti minimi
L. Extensor digitorum

ANSWERS

1.	E	36.	C	71.	C
2.	D	37.	D	72.	D
3.	E	38.	B	73.	E
4.	E	39.	B	74.	B
5.	A	40.	D	75.	D
6.	A	41.	B	76.	C
7.	D	42.	A	77.	A
8.	B	43.	B	78.	E
9.	E	44.	D	79.	B
10.	B	45.	B	80.	B
11.	E	46.	D	81.	D
12.	B	47.	C	82.	A
13.	B	48.	D	83.	E
14.	A	49.	A	84.	C
15.	A	50.	A	85.	B
16.	A	51.	C	86.	E
17.	A	52.	C	87.	C
18.	A	53.	A	88.	A
19.	E	54.	B	89.	D
20.	D	55.	D	90.	B
21.	B	56.	B	91	E
22.	D	57.	B	92.	A
23.	D	58.	A	93.	C
24.	B	59.	A	94.	D
25.	A	60.	D	95.	F
26.	E	61.	C	96.	F
27.	E	62.	A	97.	F
28.	A	63.	D	98.	F
29.	E	64.	A	99.	T
30.	A	65.	D	100.	F
31.	D	66.	B	101.	F
32.	A	67.	C	102.	F
33.	B	68.	D	103.	T
34.	B	69.	C	104.	T
35.	A	70.	A	105.	F

106.	F	127.	T	148.	T
107.	F	128.	T	149.	F
108.	F	129.	T	150.	H
109.	T	130.	T	151.	L
110.	T	131.	F	152.	A
111.	F	132.	T	153.	F
112.	T	133.	T	154.	I
113.	F	134.	F	155.	K
114.	T	135.	F	156.	B
115.	T	136.	F	157.	D
116.	T	137.	F	158.	L
117.	T	138.	T	159.	C
118.	F	139.	F	160.	F
119.	T	140.	F	161.	N
120.	T	141.	T	162.	M
121.	F	142.	F	163.	K
122.	F	143.	T	164.	G
123.	T	144.	T	165.	E
124.	F	145.	T	166.	I
125.	F	146.	T	167.	L
126.	T	147.	T	168.	K
				169.	H

SECTION II — LOWER LIMB

QUESTIONS 1–42:

FOR EACH OF THE FOLLOWING MULTIPLE CHOICE QUESTIONS SELECT THE *ONE* MOST APPROPRIATE ANSWER:

1. THE SUPERFICIAL INGUINAL NODES RECEIVE LYMPH FROM:
   A. Lower abdominal wall
   B. Perineum
   C. Skin between the toes
   D. All of the above
   E. A and B only

2. THE QUADRICEPS FEMORIS MUSCLE:
   A. Extends the knee
   B. Flexes the knee
   C. Extends the hip
   D. Rotates the knee
   E. Abducts the knee

3. THE FEMORAL TRIANGLE IS BOUNDED BY:
   A. Inguinal ligament, pectineus, sartorius
   B. Inguinal ligament, adductor longus, gracilis
   C. Inguinal ligament, rectus femoris, sartorius
   D. Inguinal ligament, adductor longus, sartorius
   E. None of the above

4. THE FEMORAL SHEATH IS OCCUPIED BY:
   A. Femoral artery
   B. Femoral vein
   C. Femoral nerve
   D. All of the above
   E. A and B only

5. THE FEMORAL RING:
   A. Is medial to femoral vein
   B. Is lateral to lacunar ligament
   C. Is posterior to inguinal ligament
   D. All of the above
   E. A and B only

6. ARTERY IN ADDUCTOR CANAL:
   A. Femoral
   B. Obturator
   C. Profunda femoris
   D. Perforating branch of profunda femoris
   E. Medial circumflex femoral

7. THE ORIFICE IN ADDUCTOR MAGNUS MUSCLE TRANSMITS:
   A. Femoral vessels
   B. Femoral nerve
   C. Saphenous nerve
   D. Tibial nerve
   E. Sciatic nerve

8. A TRIBUTARY OF THE LONG SAPHENOUS VEIN:
   A. Short saphenous
   B. Sural
   C. Superficial epigastric
   D. Femoral
   E. Popliteal

9. THE GREATER SCIATIC FORAMEN TRANSMITS THE FOLLOWING STRUCTURES, *EXCEPT:*
   A. Superior gluteal vessels
   B. Posterior cutaneous nerve of thigh
   C. Piriformis muscle
   D. Obturator internus
   E. Inferior gluteal vessels

10. THE FOLLOWING MUSCLES ARE INSERTED INTO THE GREATER TROCHANTER OF FEMUR, *EXCEPT:*
    A. Gluteus maximus
    B. Gluteus medius
    C. Gluteus minimus
    D. Piriformis
    E. Obturator externus

11. MUSCLE PAIR INSERTED INTO ILIOTIBIAL TRACT:
    A. Gluteus maximus and gluteus medius
    B. Gluteus medius and gluteus minimus
    C. Quadratus femoris and gluteus maximus
    D. Tensor fasciae latae and quadratus femoris
    E. Tensor fasciae latae and gluteus maximus

12. TO AVOID THE SCIATIC NERVE, AN INJECTION INTO THE BUTTOCK IS BEST GIVEN INTO:
    A. Upper and outer quadrant
    B. Upper and inner quadrant
    C. Lower and inner quadrant
    D. Lower and outer quadrant
    E. At the junction of the four quadrants

13. THE SCIATIC NERVE SUPPLIES THE FOLLOWING MUSCLES *EXCEPT:*
    A. Biceps femoris
    B. Semitendinosus
    C. Semimembranosus
    D. Gluteus maximus
    E. Adductor magnus

14. IN THE HIP JOINT THE SYNOVIAL MEMBRANE DOES *NOT* LINE THE:
    A. Inner surface of the capsule
    B. Ligament of the head of the femur
    C. Articular cartilage
    D. Non-articular surface of the femur
    E. Acetabular pad of fat

15. FLEXION OF THE HIP JOINT IS CARRIED OUT BY:
    A. Iliopsoas
    B. Vastus intermedius
    C. Semimembranosus
    D. Gluteus maximus
    E. Quadratus femoris

16. BONY PROMINENCES ON WHICH ONE NORMALLY KNEELS:
    A. Femoral condyles
    B. Patellae
    C. Tibial condyles
    D. Intercondylar eminences of tibia
    E. Tibial tuberosities

17. MUSCLE WHICH FLEXES HIP *AND* KNEE:
    A. Rectus femoris
    B. Semitendinosus
    C. Biceps femoris
    D. Sartorius
    E. None of the above

18. THE STRUCTURE CLOSEST TO THE POSTERIOR LIGAMENT OF THE KNEE JOINT:
    A. Popliteal artery
    B. Popliteal vein
    C. Tibial nerve
    D. Common peroneal nerve
    E. Sural nerve

19. LIGAMENT(S) TAUT IN FULL EXTENSION OF THE KNEE:
    A. Collateral
    B. Anterior cruciate
    C. Oblique popliteal
    D. All of the above
    E. A and B only

20. THE FOLLOWING BURSA INVARIABLY COMMUNICATES WITH THE KNEE JOINT:
    A. Suprapatellar
    B. Prepatellar
    C. Subcutaneous infrapatellar
    D. Deep infrapatellar
    E. Semimembranosus

21. THE SAPHENOUS NERVE:
    A. Is a branch of the obturator
    B. Gives a branch to the scrotum
    C. Is closely related to the great saphenous vein in the upper thigh
    D. Is cutaneous to the medial side of the foot
    E. Is motor to adductor magnus

22. THE CORTEX OF THE SHAFT OF THE TIBIA RECEIVES ITS MAIN BLOOD SUPPLY FROM:
    A. Arteries in attached muscles
    B. The collateral circulation at knee and ankle
    C. The nutrient artery
    D. The circulus vasculosus
    E. None of the above

23. THE EPIPHYSIS OF THE UPPER END OF TIBIA INCLUDES THE ATTACHMENT OF:
    A. Ligamentum patellae
    B. Fibular collateral ligament
    C. Sartorius
    D. Popliteus
    E. Gracilis

24. THE SUPERFICIAL PERONEAL NERVE SUPPLIES:
    A. Peroneus longus and brevis
    B. Peroneus tertius
    C. Tibialis anterior
    D. Extensor digitorum longus
    E. Flexor digitorum longus

25. THE DEEP PERONEAL NERVE SUPPLIES:
    A. Tibialis anterior
    B. Extensor hallucis longus
    C. Extensor digitorum longus
    D. All of the above
    E. A and B only

26. STRUCTURE CLOSEST TO THE SKIN OF THE SOLE:
    A. Flexor digitorum brevis
    B. Quadratus plantae
    C. Plantar aponeurosis
    D. Long plantar ligament
    E. Short plantar ligament

27. SESAMOID BONES IN THE FOOT ARE FOUND IN:
    A. Flexor hallucis longus
    B. Flexor hallucis brevis
    C. Flexor digitorum longus
    D. Flexor digitorum brevis
    E. Quadratus plantae

28. THE TALUS PROVIDES ORIGIN FOR:
    A. Extensor digitorum brevis
    B. Quadratus plantae
    C. Flexor hallucis longus
    D. Flexor digitorum brevis
    E. None of the above

29. THE MEDIAL PLANTAR NERVE:
   A. Supplies cutaneous branches to 3½ toes
   B. Supplies motor branches to the interossei
   C. Supplies adductor hallucis
   D. Is homologous with the ulnar nerve in the hand
   E. Passes through the first intermetatarsal space

30. THE 'KEYSTONE' OF THE MEDIAL LONGITUDINAL ARCH OF THE FOOT IS:
   A. Talus
   B. Calcaneus
   C. Navicular
   D. Cuboid
   E. First metatarsal

31. THE MEDIAL LONGITUDINAL ARCH OF THE FOOT IS SUPPORTED BY:
   A. Tibialis posterior
   B. Flexor hallucis longus
   C. The 'spring' ligament
   D. The plantar aponeurosis
   E. All of the above

32. INVERSION OF THE FOOT IS PERFORMED BY:
   A. Peroneus longus and brevis
   B. Peroneus longus and tibialis posterior
   C. Tibialis anterior and tibialis posterior
   D. Tibialis anterior and peroneus tertius
   E. None of the above muscle pairs

33. THE PLANTAR NERVES AND VESSELS LIE BETWEEN:
   A. Plantar aponeurosis and the first muscle layer
   B. First and second muscle layers
   C. Second and third muscle layers
   D. Third and fourth muscle layers
   E. Plantar aponeurosis and skin

34. THE CUBOID IS GROOVED BY THE TENDON OF:
    A. Peroneus longus
    B. Peroneus brevis
    C. Peroneus tertius
    D. Tibialis posterior
    E. Flexor hallucis longus

35. THE ANKLE JOINT HAS GREATEST FREEDOM OF MOVEMENT WHEN:
    A. It is plantar flexed
    B. It is dorsi-flexed
    C. The foot is inverted
    D. The foot is everted
    E. The forefoot is adducted

36. THE INVERTED POSTURE OF THE NEWBORN FOOT IS DUE TO:
    A. Angulation of the neck of the talus
    B. Ligamentous shortening at the subtalar joint
    C. Contraction of tibialis anterior and posterior muscles
    D. Torsion at the talonavicular joint
    E. Late development of the Achilles tendon

37. THE DORSALIS PEDIS ARTERY TERMINATES BY:
    A. Dividing in the cleft between great and second toes
    B. Joining the plantar arch
    C. Forming a dorsal arterial arch
    D. Dividing into medial and lateral plantar arteries
    E. Supplying the ankle joint

38. NUMBER OF TARSAL BONES ARTICULATING WITH THE FIVE METATARSALS:
    A. Two
    B. Three
    C. Four
    D. Five
    E. Seven

39. THE MID-TARSAL JOINT:
   A. Is between talus and calcaneus
   B. Is between talus and navicular
   C. Comprises the talonavicular and calcaneo-cuboid joints
   D. Is a purely fibrous joint
   E. Permits dorsi-flexion of the foot

40. THE EARLIEST EPIPHYSEAL CENTRE OF OSSIFICATION TO APPEAR IN THE LOWER LIMB IS FOR THE:
   A. Lower end of femur
   B. Upper end of tibia
   C. Calcaneus
   D. Patella
   E. Head of fibula

41. THE FOLLOWING MUSCLE OF THE HAND HAS NO HOMOLOGUE IN THE FOOT:
   A. Abductor pollicis brevis
   B. Adductor pollicis
   C. Opponens pollicis
   D. First dorsal interosseous
   E. First lumbrical

42. THE FAILURE OF BOTH PAIRS OF APPENDAGES OF THE HUMAN EMBRYO TO DEVELOP RESULTS IN THE DEFORMITY KNOWN AS:
   A. Polydactyly
   B. Oligodactyly
   C. Amelia
   D. Sirenomelia
   E. Phocomelia

QUESTIONS 43–70:

THE SET OF LETTERED HEADINGS BELOW IS FOLLOWED BY A LIST OF NUMBERED WORDS OR PHRASES. FOR EACH NUMBERED WORD OR PHRASE SELECT THE CORRECT ANSWER UNDER:

    A.   If the item is associated with A only
    B.   If the item is associated with B only
    C.   If the item is associated with both A and B
    D.   If the item is associated with neither A nor B

    A.   Acetabulum
    B.   Head of femur
    C.   Both
    D.   Neither

43. Articulating surface devoid of hyaline cartilage
44. Ligament attached near centre
45. Joint capsule attached to margin
46. Centre of ossification appears in first postnatal year

    A.   Vastus medialis
    B.   Vastus lateralis
    C.   Both
    D.   Neither

47. Partial origin from hip bone
48. Origin from linea aspera
49. Muscular down to the level of the patella
50. Supplied only by femoral nerve

    A.   Long saphenous vein
    B.   Short saphenous vein
    C.   Both
    D.   Neither

51. Anterior to medial malleolus
52. Related to saphenous nerve in the leg
53. Characterised by absence of valves
54. Perforating branches to deep veins of leg

A. Gluteus maximus
B. Gluteus medius
C. Both
D. Neither

55. Active in abduction of hip
56. Active in pelvic support when opposite limb is off the ground
57. Stabilising action on the knee joint
58. Supplied by the superior gluteal nerve

A. Semitendinosus
B. Biceps femoris
C. Both
D. Neither

59. Origin exclusively from ischial tuberosity
60. Insertion into head of fibula
61. Medial rotator of flexed leg
62. Supplied by obturator nerve

A. Tibia
B. Fibula
C. Both
D. Neither

63. Articulation with lateral femoral condyle
64. Articulation with body of talus
65. Gives origin to tibialis anterior
66. Upper epiphysis appears before lower epiphysis

A. Medial plantar nerve
B. Lateral plantar nerve
C. Both
D. Neither

67. Derived from common peroneal nerve
68. Cutaneous to sole of foot
69. Motor to long flexor muscles
70. Motor to extensor digitorum brevis

QUESTIONS 71–82:

DIRECTIONS: In the following series of questions, one or more of the four items is/are correct. Answer A if 1, 2, 3 are correct; B if 1 and 3 are correct; C if 2 and 4 are correct; D if only 4 is correct; and E if all four are correct.

71. THE LIGAMENT OF THE HEAD OF FEMUR:
    1. Is attached to the margins of the acetabular notch
    2. Conveys blood vessels to the head of femur
    3. Is invested by synovial membrane
    4. Is sometimes absent

72. THE RECTUS FEMORIS MUSCLE:
    1. Is a flexor of the hip joint
    2. Is supplied by the femoral nerve
    3. Arises by two heads
    4. Inserts into the midshaft of femur

73. THE BICEPS FEMORIS MUSCLE:
    1. Has an origin from the ischial tuberosity
    2. Inserts into head of fibula
    3. Is a flexor of the knee joint
    4. Is a lateral rotator of the knee joint

74. THE ADDUCTOR TUBERCLE:
    1. Gives origin to the popliteus
    2. Receives the tendon of adductor magnus
    3. Gives attachment to the medial semilunar cartilage
    4. Is at the level of an epiphyseal plate

75. A PATIENT HAS SUFFERED FRACTURE OF THE NECK OF FIBULA. INVOLVEMENT OF THE COMMON PERONEAL NERVE IN THE INJURY WOULD BE INDICATED BY:
    1. Inability to dorsi-flex the ankle
    2. Inability to plantarflex the ankle
    3. Anaesthesia of the dorsum of the foot
    4. Anaesthesia of the sole

76. TIBIALIS ANTERIOR:
    1. Arises from the lateral surface of tibia
    2. Inserts into the medial surface of the medial cuneiform bone
    3. Passes beneath the extensor retinacula of the ankle
    4. Is invested by a synovial sheath at the ankle

77. THE GASTROCNEMIUS IS:
    1. Portion of the triceps surae
    2. Supplied by the tibial nerve
    3. A flexor of the knee
    4. A flexor of the ankle

78. THE COMMON PERONEAL NERVE:
    1. Supplies the long head of biceps femoris
    2. Forms the lateral plantar nerve
    3. Passes between tibia and fibula
    4. Is a branch of the sciatic nerve

79. BONES ARTICULATING WITH THE NAVICULAR:
    1. Talus
    2. Intermediate cuneiform
    3. Lateral cuneiform
    4. Calcaneus

80. THE DORSALIS PEDIS ARTERY:
    1. Is the continuation of the anterior tibial
    2. Is palpable between tibialis anterior and flexor hallucis longus tendons
    3. Passes between the first and second metatarsal bones
    4. Is the main blood supply to the sole of the foot

81. EXTENSOR DIGITORUM BREVIS IS:
    1. Inserted into the lateral four toes
    2. Supplied by the superficial peroneal nerve
    3. An important support for the medial longitudinal arch of the foot
    4. Crossed superficially by the tendons of extensor digitorum longus

82. THE DELTOID LIGAMENT IS ATTACHED TO THE:
   1. Medial malleolus
   2. Sustentaculum tali
   3. 'Spring' ligament
   4. Tubercle of the navicular

QUESTIONS 83–100:

THE GROUP OF QUESTIONS BELOW CONSISTS OF NUMBERED HEADINGS, FOLLOWED BY A LIST OF LETTERED WORDS OR PHRASES. FOR EACH HEADING SELECT THE *ONE* WORD OR PHRASE WHICH IS MOST CLOSELY RELATED TO IT.
NOTE: EACH CHOICE MAY BE USED *ONLY ONCE*.

83. Femoral artery
84. Popliteal artery
85. Profunda femoris artery
86. External iliac artery

A. Inferior epigastric artery
B. Superficial epigastric artery
C. Perforating arteries
D. Paired genicular arteries

87. Intertrochanteric line
88. Ischial spine
89. Ischial tuberosity
90. Intertrochanteric crest
91. Pubic tubercle

A. Sacrotuberous ligament
B. Iliofemoral ligament
C. Sacrospinous ligament
D. Inguinal ligament
E. None of the above

92. Adductor magnus
93. Biceps femoris
94. Gracilis
95. Vastus intermedius
96. Tensor fasciae latae

A. Femoral nerve
B. Obturator nerve
C. Sciatic nerve
D. Superior gluteal nerve
E. Two of the above nerves

97. Peroneus longus
98. Peroneus tertius
99. Flexor hallucis longus
100. Tibialis posterior

A. Groove under sustentaculum tali
B. Fifth metatarsal
C. Navicular
D. Groove under cuboid

QUESTIONS 101–170:

IN REPLY TO THE FOLLOWING QUESTIONS INDICATE WHETHER YOU THINK EACH STATEMENT IS *TRUE* OR *FALSE:*

THE FEMORAL NERVE:
101. Arises from the upper sacral nerves
102. Leaves pelvis via femoral ring
103. Supplies the main flexor muscle of the hip joint
104. Gives a cutaneous branch to leg and foot
105. Has no articular branch

THE FEMORAL SHEATH:
106. Is formed by the fascia transversalis and the fascia iliaca
107. Surrounds both the femoral artery and the femoral nerve
108. Contains lymph nodes which drain portion of perineum
109. Is an important cause of varicose veins in man as it prevents dilatation of the femoral vein during periods of increased blood flow
110. Opens at the fossa ovalis

THE FEMORAL VEIN:
111. Is a direct continuation of the popliteal vein
112. Proximal portion occupies femoral sheath
113. Lies medial to the femoral artery in the upper part of the femoral triangle
114. Receives the great saphenous vein directly
115. When cut, bleeds profusely because there are no competent valves between it and the heart

THE PROFUNDA FEMORIS ARTERY:
116. Generally arises from the external iliac artery
117. Is distributed to the adductor muscles
118. Anastomoses with the inferior gluteal and popliteal arteries
119. Is easily palpable in the femoral triangle
120. Because of its many branches, is easily distinguished from the femoral artery in an arteriogram

THE OBTURATOR NERVE:
121. Enters the thigh by piercing the lacunar ligament
122. Is cutaneous to medial side of thigh
123. Supplies the iliopsoas muscle
124. Gives branches to both hip and knee joints
125. Bilateral injury to the nerve causes the condition known as 'scissors gait'

THE QUADRICEPS FEMORIS:
126. Arises entirely from the femur
127. Is the main extensor of the knee joint
128. Is supplied by the femoral nerve
129. Gives fibrous expansions to reinforce capsule of knee joint
130. Is inserted via the ligamentum patellae into the upper epiphysis of the tibia

THE HIP JOINT:
131. Is a synovial, ball-and-socket joint
132. Is usually unstable at birth because ossification in the head of the femur has not commenced
133. Depends on muscular factors for its stability in the adult
134. Synovial cavity may communicate with bursa behind iliopsoas
135. Femoral neck fractures within the joint usually deprive the femoral head of its main blood supply

GLUTEUS MAXIMUS:
136. Flexes the hip joint
137. Is attached to sacrotuberous ligament
138. Inserts entirely into gluteal ridge
139. Stabilises flexed weight-bearing knee joint
140. Is supplied by the inferior gluteal nerve

THE HAMSTRING MUSCLES:
141. Arise from ischial tuberosity
142. Insert into linea aspera
143. Receive blood from the profunda femoris artery
144. Are innervated entirely by the sciatic nerve
145. Extend the hip joint during walking

THE SCIATIC NERVE:
146. Is formed entirely by sacral spinal nerves
147. Consists of two components, which pass through the greater and lesser sciatic foramina respectively
148. Passes midway between the greater trochanter and the ischial tuberosity
149. Rests on adductor magnus
150. Terminates by dividing into superficial and deep peroneal nerves

IN THE KNEE JOINT:
151. Synovial fluid can flow easily into the suprapatellar bursa
152. The medial meniscus is attached to the tibial collateral ligament
153. The lateral meniscus is not attached to the fibular collateral ligament
154. The cruciate ligaments prevent excessive movement of the tibia in an antero-posterior direction
155. Rotation is not permitted because it is a hinge joint

THE PATELLA:
156. Is a sesamoid bone
157. Straddles the knee 'joint line' during flexion
158. May have two separate centres of ossification
159. Bears the body weight in the normal kneeling position
160. Is stabilised by the action of vastus medialis

IN THE ANKLE JOINT:
161. Movement is almost restricted to dorsi-flexion and plantar flexion
162. The dorsi-flexed position is the least stable
163. Movement at the joint causes movement in the superior tibio-fibular joint
164. The medial or deltoid ligament is weak
165. A 'sprain' commonly involves the anterior talofibular ligament

IN THE DEVELOPING LIMBS:
166. The greatest number of mitotic figures is found in the apical ectodermal ridge
167. The muscles develop *in situ* within the limb mesoderm
168. The interdigital clefts are formed by local degeneration of ectoderm and mesoderm
169. The skeleton passes through mesenchymal, cartilaginous and bony stages of development
170. Vulnerability to teratogenic agents is maximal in the 5th–8th weeks of gestation

SECTION OF GROIN

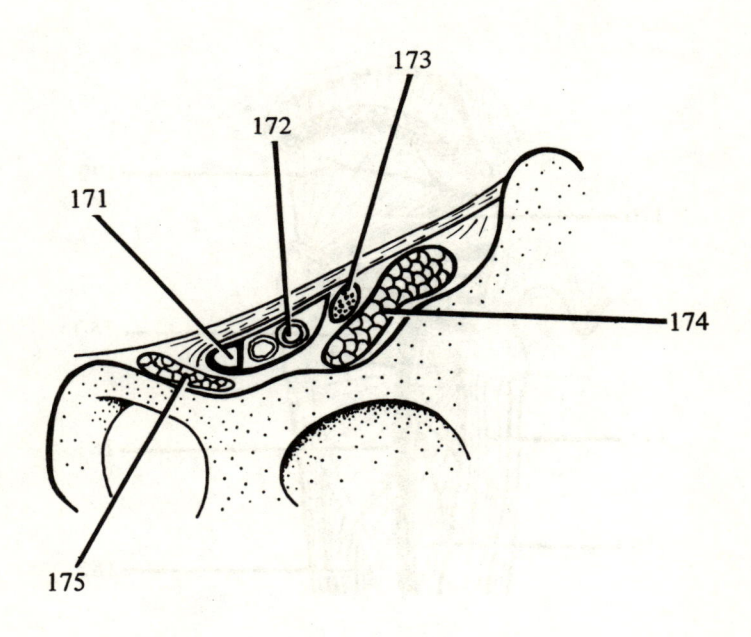

IDENTIFY THE NUMBERED STRUCTURES:

- A. Pectineus
- B. Iliopsoas
- C. Lacunar ligament
- D. Femoral vein
- E. Femoral triangle
- F. Femoral artery
- G. Femoral nerve
- H. Femoral ring
- I. Femoral sheath
- J. Pectineal ligament

GLUTEAL REGION AND THIGH

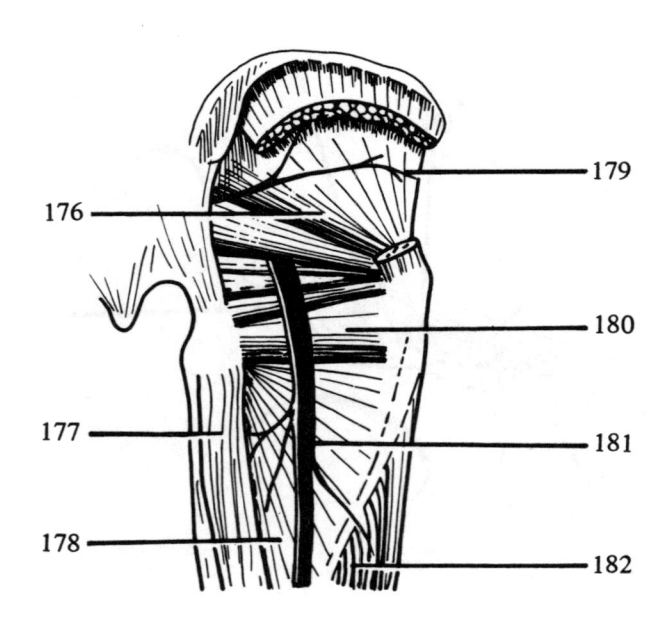

IDENTIFY THE NUMBERED STRUCTURES:

- A. Gluteus maximus
- B. Adductor magnus
- C. Gluteus medius
- D. Inferior gluteal nerve
- E. Gluteus minimus
- F. Sciatic nerve
- G. Piriformis
- H. Superior gluteal nerve
- I. Quadratus femoris
- J. Semitendinosus
- K. Obturator internus
- L. Biceps, short head

LIGAMENTS AND MUSCLES AT THE KNEE

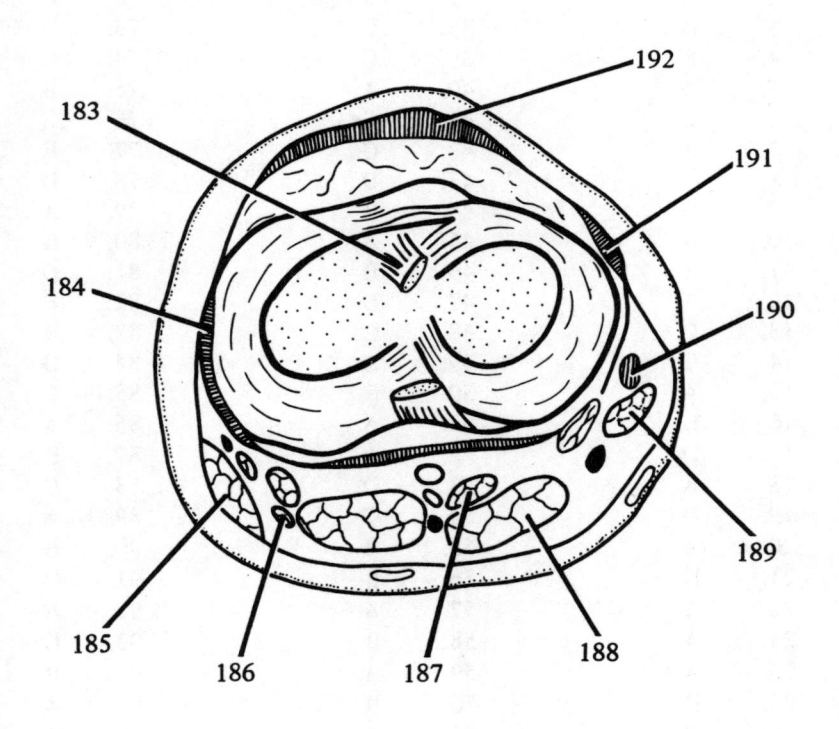

IDENTIFY THE NUMBERED STRUCTURES:

A. Sartorius
B. Plantaris
C. Semitendinosus
D. Medial meniscus
E. Soleus
F. Biceps femoris
G. Popliteus
H. Ligamentum teres

I. Fibular collateral ligament
J. Gastrocnemius
K. Posterior cruciate ligament
L. Iliotibial tract
M. Anterior cruciate ligament
N. Tendon of quadriceps
O. Tibial collateral ligament
P. Ligamentum patellae

43

ANSWERS

1.	D	36.	A	71.	E
2.	A	37.	B	72.	A
3.	D	38.	C	73.	E
4.	E	39.	C	74.	C
5.	D	40.	A	75.	B
6.	A	41.	C	76.	E
7.	A	42.	C	77.	E
8.	C	43.	D	78.	D
9.	D	44.	B	79.	A
10.	A	45.	A	80.	B
11.	E	46.	B	81.	D
12.	A	47.	D	82.	E
13.	D	48.	C	83.	B
14.	C	49.	A	84.	D
15.	A	50.	C	85.	C
16.	E	51.	A	86.	A
17.	D	52.	A	87.	B
18.	A	53.	D	88.	C
19.	D	54.	C	89.	A
20.	A	55.	B	90.	E
21.	D	56.	B	91.	D
22.	A	57.	A	92.	E
23.	A	58.	B	93.	C
24.	A	59.	A	94.	B
25.	D	60.	B	95.	A
26.	C	61.	A	96.	D
27.	B	62.	D	97.	D
28.	E	63.	A	98.	B
29.	A	64.	C	99.	A
30.	A	65.	A	100.	C
31.	E	66.	A	101.	F
32.	C	67.	D	102.	F
33.	B	68.	C	103.	F
34.	A	69.	D	104.	T
35.	A	70.	D	105.	F

106.	T	135.	T	164.	F
107.	F	136.	F	165.	T
108.	T	137.	T	166.	F
109.	F	138.	F	167.	T
110.	F	139.	T	168.	T
111.	T	140.	T	169.	T
112.	T	141.	T	170.	T
113.	T	142.	F	171.	H
114.	T	143.	T	172.	F
115.	T	144.	T	173.	G
116.	F	145.	T	174.	B
117.	T	146.	F	175.	A
118.	T	147.	F	176.	E
119.	F	148.	T	177.	J
120.	T	149.	T	178.	B
121.	F	150.	F	179.	H
122.	T	151.	T	180.	I
123.	F	152.	T	181.	F
124.	T	153.	T	182.	L
125.	F	154.	T	183.	M
126.	F	155.	F	184.	O
127.	T	156.	T	185.	A
128.	T	157.	F	186.	C
129.	T	158.	T	187.	B
130.	T	159.	F	188.	J
131.	T	160.	T	189.	F
132.	F	161.	T	190.	I
133.	F	162.	F	191.	L
134.	T	163.	T	192.	P

SECTION III — THORAX

QUESTIONS 1-19:

FOR EACH OF THE FOLLOWING MULTIPLE CHOICE QUESTIONS SELECT THE *ONE* MOST APPROPRIATE ANSWER:

1. THE STERNUM:
    A. Is composed of four parts
    B. Gives origin to pectoralis minor
    C. Articulates with the upper ten costal cartilages
    D. Contains red marrow at its upper end only
    E. Makes a synovial joint with the clavicle

2. THE INTERCOSTAL NERVES AND VESSELS COURSE BETWEEN:
    A. Skin and deep fascia
    B. Deep fascia and external intercostals
    C. External and internal intercostals
    D. Internal and innermost intercostals
    E. Innermost intercostals and endothoracic fascia

3. THE FOLLOWING RIBS ARE TYPICAL *EXCEPT* THE:
    A. First
    B. Third
    C. Fifth
    D. Seventh
    E. Ninth

4. THE MOST SUPERFICIAL STRUCTURE IN THE SUPERIOR MEDIASTINUM:
   A. Thymus
   B. Arch of aorta
   C. Left brachiocephalic vein
   D. Brachiocephalic trunk
   E. Vagus nerve

5. THE BRACHIOCEPHALIC TRUNK DIVIDES INTO TWO ARTERIES:
   A. Right and left common carotid
   B. Right common carotid and right subclavian
   C. Left common carotid and left subclavian
   D. Right and left subclavian
   E. Right and left coronary

6. THE AZYGOS VEIN:
   A. Carries blood from the oesophagus
   B. Arches directly over the inferior lobe bronchus
   C. Passes between the vagus and the trachea
   D. Joins the superior vena cava behind the first costal cartilage
   E. None of the above

7. THE PARIETAL PLEURA IS DERIVED FROM THE:
   A. Somatopleure
   B. Splanchnopleure
   C. Neural crest
   D. Foregut endoderm
   E. None of the above

8. THE NUMBER OF BRONCHO-PULMONARY SEGMENTS IN THE MIDDLE LOBE OF THE RIGHT LUNG IS:
   A. One
   B. Two
   C. Three
   D. Four
   E. From two to four

9. HEART VALVES ARE:
   A. Entirely avascular
   B. Designed to be two-way valves
   C. Composed mainly of muscle
   D. Contains fibro-elastic tissue
   E. Rarely affected by disease

10. THE NUMBER OF PULMONARY VEINS ENTERING THE LEFT ATRIUM IS NORMALLY:
    A. One
    B. Two
    C. Four
    D. Six
    E. Eight

11. THE BUNDLE OF HIS USUALLY RECEIVES ITS BLOOD SUPPLY FROM THE:
    A. Interventricular branch of the right coronary artery
    B. Interventricular branch of the left coronary artery
    C. Right marginal artery
    D. Left marginal artery
    E. Left coronary trunk directly

12. CONDUCTING TISSUE IN THE HEART IS COMPOSED OF:
    A. Connective tissue
    B. Modified nerves
    C. Modified heart muscle
    D. Smooth muscle
    E. Sinusoidal blood vessels

13. THE FORAMEN OVALE IN THE FETAL HEART:
    A. Conducts blood from right to left atrium
    B. Is created by the septum secundum
    C. Is protected by the septum primum
    D. All of the above
    E. None of the above

14. THE FOLLOWING SURFACE LANDMARK IS A GUIDE TO THE GASTRO-OESOPHAGEAL ORIFICE:
    A. Seventh left costal cartilage
    B. Left linea semilunaris
    C. Tip of the ninth left costal cartilage
    D. Xiphisternal joint
    E. Left nipple

15. THE OESOPHAGEAL LUMEN IS:
    A. Connected to pleural cavity in the embryo
    B. Lined by ciliated pseudo-stratified columnar epithelium in the adult
    C. Slit-like except during passage of a bolus of food
    D. All of the above
    E. A and B only

16. THE CENTRAL TENDON OF THE DIAPHRAGM:
    A. Is derived from the septum transversum
    B. Is attached to the parietal pericardium
    C. Is pierced by the oesophagus
    D. All of the above
    E. A and B only

17. THE ASCENDING AORTA:
    A. Has no branches
    B. Begins at the semilunar valves
    C. Arises from the right ventricle
    D. Occupies the superior mediastinum
    E. Is a highly muscular artery

18. THE LIGAMENTUM ARTERIOSUM IS DERIVED FROM THE:
    A. Ductus arteriosus
    B. Glomus arteriosus
    C. Conus arteriosus
    D. Truncus arteriosus
    E. Umbilical artery

19. THE WALL OF THE AORTA IS NOTABLE FOR ITS HIGH CONTENT OF:
   A. Muscle cells
   B. Elastic fibres
   C. Collagen fibres
   D. Muscle and collagen
   E. Muscle and elastic fibres

QUESTIONS 20—44:

THE SET OF LETTERED HEADINGS BELOW IS FOLLOWED BY A LIST OF NUMBERED WORDS OR PHRASES. FOR EACH NUMBERED WORD OR PHRASE SELECT THE CORRECT ANSWER UNDER:

   A. If the item is associated with A only
   B. If the item is associated with B only
   C. If the item is associated with both A and B
   D. If the item is associated with neither A nor B

   A. Right lung
   B. Left lung
   C. Both
   D. Neither

20. Attached by ligaments to chest wall
21. An oblique fissure is present
22. Impressed by azygos vein
23. Four broncho-pulmonary segments present in the upper lobe
24. Related to lower thoracic oesophagus

A. Left bronchus
B. Right bronchus
C. Both
D. Neither

25. Origin from a diverticulum at distal end of laryngotracheal groove
26. Approximately 2.5 cm. long, forming an angle of 25° with the trachea
27. Lined by pseudo-stratified ciliated columnar epithelium
28. Directly related to the oesophagus
29. Divides into three secondary (lobar) bronchi

A. Right ventricle
B. Left ventricle
C. Both
D. Neither

30. Develops entirely from the primitive embryonic ventricle
31. Forms part of diaphragmatic surface of heart
32. Contains the atrioventricular node in its wall
33. Frequently contains the septomarginal trabecula
34. Mitral valve guards its atrioventricular opening

A. Sinuatrial node
B. Atrioventricular node
C. Both
D. Neither

35. Develops from neural crest
36. Lies close to superior vena cava
37. Initiates the heart beat
38. Differs histologically from cardiac muscle
39. Blood supply from right coronary artery

A. Left coronary artery
B. Right coronary artery
C. Both
D. Neither

40. Origin from pulmonary artery
41. Main source of blood supply to bundle of His
42. Main blood supply to left ventricle
43. Gives off posterior interventricular artery
44. Origin from an aortic sinus

QUESTIONS 45–50:

DIRECTIONS: In the following series of questions, one or more of the four items is/are correct. Answer A if 1, 2, 3 are correct; B if 1 and 3 are correct; C if 2 and 4 are correct; D if only 4 is correct; and E if all four are correct.

45. THE LOWER APERTURE OF THE THORAX IS FORMED BY THE:
    1. Lower costal cartilages
    2. Twelfth rib
    3. Xiphoid process
    4. Body of twelfth thoracic vertebra

46. AMONG THE CONTENTS OF THE MEDIASTINUM:
    1. Trachea
    2. Lungs
    3. Phrenic nerves
    4. Diaphragm

47. THE FOLLOWING MAY IMPRESS THE OESOPHAGUS DURING PASSAGE OF A BARIUM SWALLOW:
    1. Arch of aorta
    2. Left bronchus
    3. Left atrium
    4. Left ventricle

48. THE THORACIC DUCT RETURNS LYMPH FROM ALL AREAS OF THE BODY *EXCEPT:*
    1. Left half of the body below the diaphragm
    2. Right half of the body below the diaphragm
    3. Left half of the body above the diaphragm
    4. Right half of the body above the diaphragm

49. THE DEVELOPING DIAPHRAGM RECEIVES CONTRIBUTIONS FROM THE:
    1. Septum transversum
    2. Ventral mesogastrium
    3. Pleuroperitoneal membranes
    4. Dorsal mesocardium

50. A 3-INCH STAB WOUND PASSING RADIALLY THROUGH THE LEFT NINTH INTERCOSTAL SPACE WILL INVOLVE THE:
    1. Spleen
    2. Lung
    3. Pleura
    4. Kidney

QUESTIONS 51–145:

IN REPLY TO THE FOLLOWING QUESTIONS INDICATE WHETHER YOU THINK EACH STATEMENT IS *TRUE* OR *FALSE:*

THE BREAST:
51. Is connected to skin and deep fascia by fibrous bands
52. Lymphatics drain entirely into the axilla
53. Frequently extends a 'tail' into the axilla
54. Left nipple is usually a little lower than right nipple
55. Accessory nipples may appear on the abdominal skin

THE BREAST:
56. Contains about a dozen compound exocrine glands
57. Increases in size at puberty, mainly due to the accumulation of fat in the connective tissue
58. Has lactiferous ducts lined by two layers of columnar epithelium
59. Secretes milk under the influence of the adrenal cortex
60. Contains myoepithelial cells which surround the alveoli and aid in the expression of milk

THE THYMUS:
61. Increases in size until puberty
62. Has more lymphocytes in its efferent than in its afferent blood vessels
63. Has Hassall's corpuscles, formed from reticular epithelial cells, in its medulla
64. Is necessary in early life for the proper development of lymphoid tissue elsewhere in the body
65. Has little or no concern with the development of plasma cells

THE PHRENIC NERVES:
66. Arise from the cervical plexus C3, 4, 5
67. Supply the skin over the deltoid muscles
68. Enter the thorax by crossing the necks of the first ribs
69. Give sensory branches to the mediastinal and diaphragmatic pleura
70. Injury results in inability to breathe

THE BRACHIOCEPHALIC VEINS:
71. Possess valves which prevent reverse blood flow during changes of intrathoracic pressure
72. Receive inferior thyroid veins
73. Lie in front of thymus when this is fully developed
74. In the infant, the left vein may encroach above the suprasternal notch
75. The right vein marks the right lung

THE LEFT VAGUS NERVE:
76. Makes up part of the cranial parasympathetic outflow
77. Contributes to both the cardiac and pulmonary plexuses
78. Is in direct contact with the aortic arch
79. Injury in the region of the aortic arch may cause hoarseness
80. Enters the abdomen as the left gastric nerve

THE LUNGS:
81. Occupy the pleural cavities in the thorax
82. Increase in length, breadth and depth during inspiration
83. Contain elastic tissue in the wall of the bronchial tree and alveoli
84. Have intrapulmonary bronchioles with cartilage bars
85. Have no lymphatic drainage

THE RIGHT LUNG:
86. Due to its greater size, its bronchus is longer than the left bronchus
87. Auscultation of the middle lobe is best carried out immediately inferior to the scapula
88. The bronchus to the apical segment of the lower lobe is the first posterior branch of the bronchial tree
89. Expansion is brought about by muscles in the bronchial tree
90. Expiration is a passive movement due to the elastic recoil of the lung

THE CORONARY ARTERIES:
91. Arise from the aortic sinuses
92. Are unusual in that they fill mainly in diastole
93. The right artery generally supplies the atrioventricular node
94. The left artery runs in the interventricular groove with the great cardiac vein
95. In the young adult the anastomoses between the arterioles are sufficient to prevent heart damage when one of the arteries is suddenly occluded

IN THE OUTFLOW CHANNELS OF THE DEVELOPING HEART:
96. The bulbus cordis gives rise to the muscular part of the right ventricle
97. The bulbar septum separates the conus of the right ventricle from the conus of the left ventricle
98. The junction of the interventricular and bulbar septa is marked by the semilunar valves
99. The aorticopulmonary septum is laid down entirely within the truncus arteriosus
100. Blood from the pulmonary trunk passes into the aortic arch through the ductus arteriosus

THE SUPERIOR VENA CAVA:
101. Develops from the right horn of the sinus venosus
102. Commences by the union of right internal jugular and subclavian veins
103. Receives the azygos vein
104. Is lodged in a groove in the upper lobe of the right lung
105. Is accompanied by the vagus nerve

THE PERICARDIUM OF THE HEART:
106. Is composed of outer serous and inner fibrous layers
107. The fibrous pericardium is attached to the inferior aspect of the right ventricle
108. Attachment to central tendon of diaphragm causes the heart to descend on inspiration
109. The layer on the surface of the heart is called the epicardium
110. Visceral and parietal layers meet around the great vessels

THE LEFT ATRIUM OF THE HEART:
111. With the exception of the auricle, is smooth-walled
112. Receives the left common cardial vein in the embryo
113. Contains oxygenated blood in adult but not in fetal life
114. Has a bicuspid valve between it and the left ventricle
115. Is the most posterior heart chamber

IN THE DEVELOPING HEART:
116. The sinus venosus contributes to the formation of both atria
117. Most of the left atrium is formed from primitive pulmonary veins
118. Blood pressure is higher in the right atrium than in the left
119. A left-to-right shunt occurs through the foramen ovale
120. Deoxygenated blood from the right atrium passes mainly into the right ventricle

THE OESOPHAGUS:
121. Is lined by columnar epithelium to facilitate absorption
122. Has few glands for lubrication as the food is already lubricated and passes through quickly
123. Has a muscularis externa entirely composed of smooth muscle
124. Makes no contact with the pleura
125. Plays no part in digestion

THE OESOPHAGUS:
126. Contains voluntary muscle in its upper portion
127. Is crossed by the left main bronchus
128. Has the left atrium as a direct anterior relation
129. Lies in front of part of the descending aorta
130. Some of its venous blood passes through the portal system

THE THORACIC DUCT:
131. Frequently commences as a dilated sac called the cisterna chyli
132. Pierces the diaphragm with the oesophagus
133. Veers sharply from right to left at the level of the intervertebral disc between T4 and T5
134. Drains most of the lymph from both the right and left lungs
135. Terminates by entering the junction of the left internal jugular and subclavian veins

INTERCOSTAL NERVES:
136. Supply both external and internal intercostal muscles
137. Supply the costal slips of origin of pectoralis major and minor
138. Supply the costal pleura
139. Each supplies skin overlying three successive intercostal spaces
140. Lower six have homologous motor and sensory functions in the anterior abdominal wall

THE DIAPHRAGM:
141. Moves upwards when it contracts
142. Muscular portion is pierced by inferior vena cava
143. Oesophagus is encircled by the right crus
144. Motor innervation is from phrenic nerves
145. The paralysed hemidiaphragm is seen to be higher than its opposite fellow on x-ray taken in inspiration

MEDIASTINUM

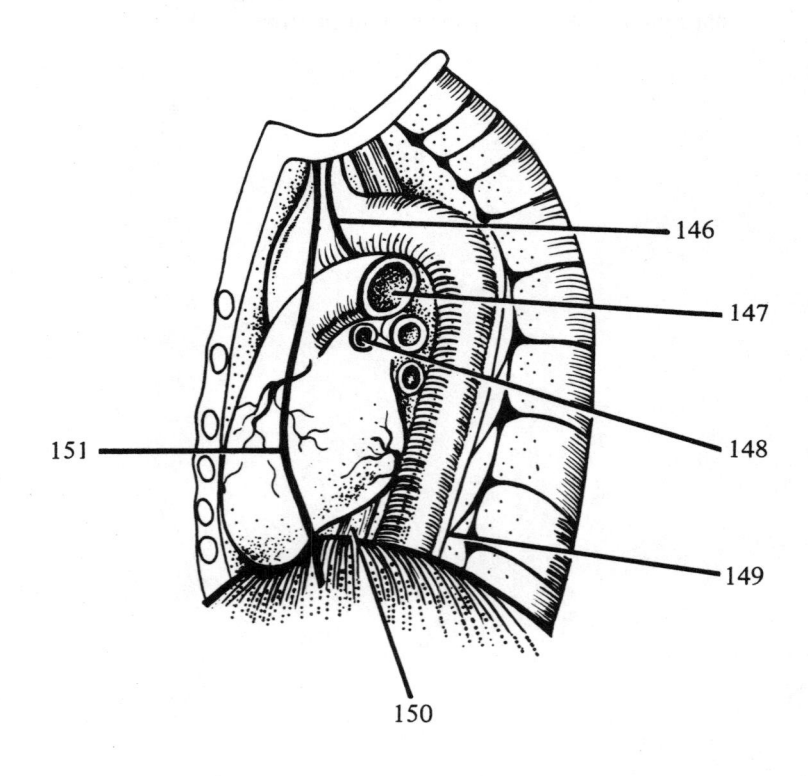

IDENTIFY THE NUMBERED STRUCTURES:

- A. Oesophagus
- B. Trachea
- C. Pulmonary vein
- D. Pulmonary artery
- E. Vagus nerve
- F. Left bronchus
- G. Greater splanchnic nerve
- H. Gastric nerve
- I. Thoracic duct
- J. Recurrent laryngeal nerve
- K. Phrenic nerve
- L. Internal thoracic artery
- M. Aortic arch

SECTION OF THORAX

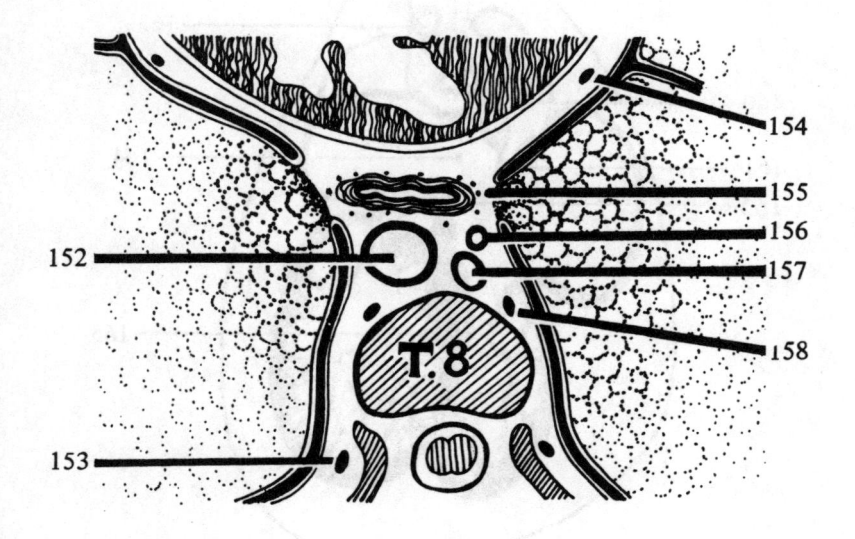

IDENTIFY THE NUMBERED STRUCTURES:

A.	Oesophageal plexus
B.	Trachea
C.	Vagus nerve
D.	Pericardium
E.	Pleura
F.	Aorta
G.	Inferior vena cava
H.	Azygos vein
I.	Phrenic nerve
J.	Thoracic duct
K.	Sympathetic chain
L.	Greater splanchnic nerve

SECTION OF THORAX, 7 WEEK EMBRYO

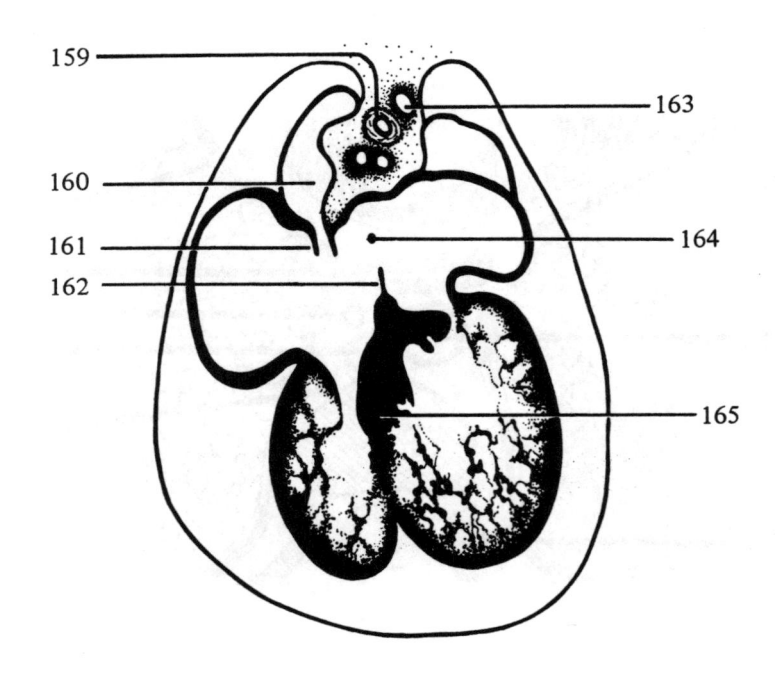

IDENTIFY THE NUMBERED STRUCTURES:

A.	Ostium primum	H.	Oesophagus
B.	Superior vena cava	I.	Septum secundum
C.	Ostium secundum	J.	Common cardinal vein
D.	Venous valve	K.	Sinus venosus
E.	Septum primum	L.	Foramen ovale
F.	Dorsal aorta	M.	Trachea
G.	Interventricular septum	N.	Bulbus cordis

ANSWERS

1.	E	36.	A	71.	F
2.	D	37.	A	72.	T
3.	A	38.	C	73.	F
4.	A	39.	C	74.	T
5.	B	40.	D	75.	T
6.	A	41.	B	76.	T
7.	A	42.	A	77.	T
8.	B	43.	B	78.	T
9.	D	44.	C	79.	T
10.	C	45.	E	80.	F
11.	A	46.	B	81.	T
12.	C	47.	A	82.	T
13.	D	48.	B	83.	T
14.	A	49.	B	84.	F
15.	C	50.	A	85.	F
16.	E	51.	T	86.	F
17.	B	52.	F	87.	F
18.	A	53.	T	88.	T
19.	B	54.	F	89.	F
20.	D	55.	T	90.	T
21.	C	56.	T	91.	T
22.	A	57.	T	92.	T
23.	D	58.	T	93.	T
24.	B	59.	F	94.	F
25.	C	60.	T	95.	F
26.	B	61.	T	96.	F
27.	C	62.	T	97.	T
28.	A	63.	T	98.	F
29.	B	64.	T	99.	T
30.	D	65.	T	100.	T
31.	C	66.	T	101.	F
32.	D	67.	F	102.	F
33.	A	68.	F	103.	T
34.	B	69.	T	104.	T
35.	D	70.	F	105.	F

106.	F	126.	T	146.	E
107.	F	127.	T	147.	D
108.	T	128.	T	148.	C
109.	T	129.	T	149.	G
110.	T	130.	T	150.	A
111.	T	131.	T	151.	K
112.	F	132.	F	152.	F
113.	F	133.	F	153.	K
114.	T	134.	F	154.	I
115.	T	135.	T	155.	A
116.	F	136.	T	156.	J
117.	T	137.	F	157.	H
118.	T	138.	T	158.	L
119.	F	139.	T	159.	H
120.	T	140.	T	160.	K
121.	F	141.	F	161.	D
122.	T	142.	F	162.	E
123.	F	143.	T	163.	F
124.	F	144.	T	164.	C
125.	T	145.	T	165.	G

SECTION IV — ABDOMEN

QUESTIONS 1–23:

FOR EACH OF THE FOLLOWING MULTIPLE CHOICE QUESTIONS SELECT THE *ONE* MOST APPROPRIATE ANSWER:

1. ENCLOSED WITHIN THE SHEATH OF THE RECTUS ABDOMINIS MUSCLE:
   A. External oblique aponeurosis
   B. Linea alba
   C. Linea semilunaris
   D. Transversalis fascia
   E. None of the above

2. THE LOWER SIX INTERCOSTAL NERVES SUPPLY:
   A. Intercostal muscles only
   B. Intercostal and abdominal muscles
   C. Intercostal and abdominal muscles and overlying skin
   D. All of the above structures, together with the subjacent parietal peritoneum
   E. All of the above structures, together with both the parietal and visceral layers of the peritoneum

3. THE INTERNAL SPERMATIC FASCIA IS DERIVED FROM THE:
   A. External oblique
   B. Internal oblique
   C. Transversus abdominis
   D. Transversalis fascia
   E. Peritoneum

4. PARIETAL OR OXYNTIC CELLS:
   A. Secrete hydrochloric acid
   B. Are found in the fundus and body of the stomach
   C. Are eosinophilic
   D. All of the above
   E. A and B only

5. THE MESENTERY OF THE APPENDIX HAS AN ATTACHMENT TO THE:
   A. Caecum
   B. Ascending colon
   C. Ileum
   D. Mesentery of ileum
   E. Posterior abdominal wall

6. THE SIGMOID COLON:
   A. Develops from distal part of midgut
   B. Has a mesentery
   C. Receives motor branches from the vagus nerve
   D. Veins drain into left common iliac vein
   E. Is fairly constant in length

7. CHIEF FUNCTION(S) OF THE COLON:
   A. Absorption of water
   B. Secretion of mucus
   C. Secretion of digestive enzymes
   D. All of the above
   E. A and B only

8. THE VERMIFORM APPENDIX IS CHARACTERIZED BY:
   A. A large number of villi
   B. A large amount of lymphoid tissue in the mucosa
   C. A prominent muscularis mucosae
   D. Taeniae coli
   E. None of the above

9. THE COMMON BILE DUCT:
   A. Lies in the free edge of the lesser omentum
   B. Is formed by the junction of the right and left hepatic ducts
   C. Lies behind the portal vein
   D. All of the above
   E. A and B only

10. A BRANCH OF THE HEPATIC ARTERY:
    A. Cystic
    B. Left gastric
    C. Splenic
    D. Gastrohepatic
    E. None of the above

11. THE LEFT RENAL VEIN:
    A. Is crossed anteriorly by the superior mesenteric artery
    B. Lies behind the left renal artery
    C. Joins the portal vein behind the pancreas
    D. Receives the inferior mesenteric vein
    E. None of the above

12. THE GREATER OMENTUM IS ATTACHED TO:
    A. Liver and stomach
    B. Stomach and jejunum
    C. Jejunum and colon
    D. Stomach and colon
    E. Liver and colon

13. THE FOLLOWING STRUCTURES ARE FOUND WITHIN THE LESSER OMENTUM, *EXCEPT*:
    A. Hepatic artery
    B. Hepatic veins
    C. Common bile duct
    D. Lymphatics
    E. Vagal nerve fibres

14. THE SUPERIOR MESENTERIC VESSELS:
    A. Are the vessels of the primitive foregut
    B. Cross the third portion of duodenum
    C. Artery arises from aorta at L3 level
    D. Vein drains into the inferior vena cava
    E. All of the above

15. THE EMBRYONIC VITELLINE VEINS GIVE RISE TO THE:
    A. Portal vein
    B. Inferior vena cava
    C. Superior and inferior mesenteric veins
    D. All of the above
    E. None of the above

16. THE EPITHELIUM LINING THE URETER IS:
    A. Stratified squamous
    B. Cuboidal
    C. Ciliated columnar
    D. Transitional
    E. None of the above

17. THE COMMONEST MALFORMATION OF THE KIDNEY IS:
    A. Horseshoe kidney
    B. Congenital cystic kidney
    C. Partial duplication
    D. Unilateral agenesis
    E. Failure of ascent

18. THE NORMAL NUMBER OF LOBES IN THE KIDNEY IS:
    A. Three
    B. Five
    C. Seven
    D. Ten
    E. Twenty

19. THE INTERMEDIATE MESODERM GIVES RISE TO:
    A. The suprarenal cortex
    B. The ovary
    C. The suprarenal medulla
    D. All of the above
    E. A and B only

20. THE MESONEPHROS IS THE PERMANENT KIDNEY OF:
    A. Sharks
    B. Amphibia
    C. All mammals
    D. Primates only
    E. Man only

21. THE TWO EMBRYONIC PORTIONS OF THE PERMANENT HUMAN KIDNEY ARE:
    A. Mesonephros and metanephros
    B. Mesonephric duct and metanephros
    C. Metanephric cap and mesonephric duct
    D. Metanephric cap and ureteric diverticulum
    E. Ureteric diverticulum and nephrogenic cord

22. THE MOST COMMON POSITION OF THE VERMIFORM APPENDIX IS:
    A. Retrocaecal
    B. Retrocolic
    C. Retroileal
    D. Pelvic
    E. Subcaecal

23. VERTEBRAE HAVING BIFID SPINOUS PROCESSES:
    A. Cervical
    B. Thoracic
    C. Lumbar
    D. Sacral
    E. Coccygeal

QUESTIONS 24–65:

THE SET OF LETTERED HEADINGS BELOW IS FOLLOWED BY A LIST OF NUMBERED WORDS OR PHRASES. FOR EACH NUMBERED WORD OR PHRASE SELECT THE CORRECT ANSWER UNDER:

    A. If the item is associated with A only
    B. If the item is associated with B only
    C. If the item is associated with both A and B
    D. If the item is associated with neither A nor B

    A. Ventral mesogastrium
    B. Dorsal mesogastrium
    C. Both
    D. Neither

24. Lesser omentum
25. Greater omentum
26. Falciform ligament
27. Gastrosplenic ligament
28. Lienorenal ligament

    A. Right hepatic vein
    B. Right renal vein
    C. Both
    D. Neither

29. Direct entry into inferior vena cava
30. Carries oxygenated blood in fetus
31. Derived from right vitelline vein
32. Direct communication with the splenic vein

A. Stomach
B. Ileum
C. Both
D. Neither

33. Mucosal epithelial cells all columnar mucus-secreting
34. Mucosal epithelial cells show abundant microvilli
35. Aggregations of lymph nodules in lamina propria
36. Glands contain Paneth cells
37. Muscle coat consists of outer longitudinal and inner circular layers

A. Duodenum
B. Jejunum
C. Both
D. Neither

38. Site of Meckel's diverticulum
39. Has a mesentery throughout entire length
40. Brünner's glands are a feature
41. Derived entirely from foregut

A. Jejunum
B. Ileum
C. Both
D. Neither

42. Myenteric plexus is lodged between muscle layers
43. Blood supply from superior mesenteric artery
44. Mesentery devoid of arterial arcades
45. Appendices epiploicae are a gross feature

A. Caecum
B. Appendix
C. Both
D. Neither

46. Lined by columnar epithelium
47. Belongs to midgut loop in embryonic life
48. Blood supply indirectly from superior mesenteric artery
49. Venous drainage mainly to inferior vena cava
50. Unique to man

A. Liver
B. Spleen
C. Both
D. Neither

51. Derived from gut endoderm
52. Venous drainage to inferior vena cava
53. Part of stomach bed
54. Arterial supply via coeliac artery
55. Invested by peritoneum of greater sac

A. Stomach
B. Spleen
C. Both
D. Neither

56. Supplied by branches of splenic artery
57. Gives attachment to gastrosplenic ligament
58. A surface is coated by peritoneum of lesser sac
59. Venous blood drains into portal vein
60. In contact with left kidney

A. Proximal convoluted tubule of kidney
B. Distal convoluted tubule of kidney
C. Both
D. Neither

61. Lined by flat epithelium
62. Free surface of epithelium packed with microvilli
63. Epithelium shows basal infoldings with many mitochondria
64. Displays the macula densa where contact is made with afferent arteriole
65. Some cells are filled with juxtaglomerular granules

QUESTIONS 66–74:

DIRECTIONS: In the following series of questions, one or more of the four items is/are correct. Answer A if 1, 2, 3 are correct; B if 1 and 3 are correct; C if 2 and 4 are correct; D if only 4 is correct; and E if all four are correct.

66. THE RECTUS ABDOMINIS IS:
    1. Attached to the ninth and tenth ribs
    2. Marked in its upper half by tendinous intersections
    3. Surrounded by the rectus sheath throughout its length
    4. A flexor of the vertebral column

67. THE EPIPLOIC FORAMEN IS BOUNDED BY THE:
    1. Lesser omentum
    2. Inferior vena cava
    3. Duodenum
    4. Quadrate lobe of liver

68. THE DORSAL MESENTERY OF THE PRIMITIVE GUT GIVES RISE TO THE:
    1. Falciform ligament
    2. Lienorenal ligament
    3. Lesser omentum
    4. Greater omentum

69. THE PORTAL TRIAD CONTAINS THE:
    1. Portal vein
    2. Hepatic artery
    3. Bile duct
    4. Lymphatic duct

70. THE PORTAL VEIN RECEIVES BLOOD FROM THE:
    1. Stomach
    2. Liver
    3. Caecum
    4. Kidneys

71. BRANCHES OF SPLENIC ARTERY:
    1. Left gastric
    2. Right gastric
    3. Right gastro-epiploic
    4. Short gastric

72. CHARACTERISTIC OF LARGE INTESTINE:
    1. Taeniae coli
    2. Appendices epiploicae
    3. Haustra coli
    4. Large lumen

73. PART(S) OF COLON HAVING A MESENTERY:
    1. Ascending
    2. Transverse
    3. Descending
    4. Sigmoid

74. DIRECT RELATIONS OF THE RIGHT KIDNEY INCLUDE:
    1. Duodenum
    2. Colon
    3. Liver
    4. Quadratus lumborum

QUESTIONS 75-104:

THE GROUP OF QUESTIONS BELOW CONSISTS OF NUMBERED HEADINGS, FOLLOWED BY A LIST OF LETTERED WORDS OR PHRASES. FOR EACH HEADING SELECT THE *ONE* WORD OR PHRASE WHICH IS MOST CLOSELY RELATED TO IT.
NOTE: EACH CHOICE MAY BE USED *ONLY ONCE.*

75.	External oblique	A.	Tunica vaginalis
76.	Internal oblique	B.	External spermatic fascia
77.	Rectus abdominis	C.	Internal spermatic fascia
78.	Transversalis fascia	D.	Cremaster
79.	Peritoneum	E.	None of the above

80.	Body of gall bladder	A.	Tail of pancreas
81.	Hilus of spleen	B.	Second stage of duodenum
82.	Hepatic flexure of colon	C.	First stage of duodenum
83.	Quadrate lobe of liver	D.	Right kidney
84.	Caudate lobe of liver	E.	None of the above

85.	Foregut	A.	Ascending colon
86.	Midgut	B.	Descending colon
87.	Hindgut	C.	Rectum
88.	Cloaca	D.	Urachus
89.	Allantois	E.	None of the above

90.	Stomach	A.	Brünner's glands
91.	Duodenum	B.	Chylomicrons
92.	Jejunum	C.	Peptic cells
93.	Ileum	D.	Profusion of goblet cells
94.	Colon	E.	Profusion of lymph follicles

95.	Left gastric vein	A.	Inferior vena cava
96.	Inferior mesenteric vein	B.	Renal vein
97.	Inferior epigastric vein	C.	Portal vein
98.	Left testicular vein	D.	Splenic vein
99.	Right ovarian vein	E.	External iliac vein

100. Apex of loop of Henle
101. Collecting ducts
102. Proximal convoluted tubule
103. Renal pelvis
104. Glomerular epithelium

A. Podocytes
B. Numerous microvilli
C. Simple squamous epithelium
D. Cuboidal epithelium
E. Transitional epithelium

QUESTIONS 105–194:

IN REPLY TO THE FOLLOWING QUESTIONS INDICATE WHETHER YOU THINK EACH STATEMENT IS *TRUE* OR *FALSE:*

THE ANTERIOR ABDOMINAL WALL:
105. Muscles contract during coughing
106. Is innervated mainly by lumbar nerves
107. Skin lymphatics above umbilicus drain to axillary nodes
108. Veins below umbilicus drain mainly into the portal system
109. Inferior attachment of membranous (Scarpa's) fascia is to the inguinal ligament

THE INGUINAL CANAL:
110. Contains the spermatic cord in the male but only fat in the female
111. Normally contains a peritoneal diverticulum
112. Is more oblique in the adult than in the newborn
113. Its superficial ring is protected by the pyramidalis muscle
114. Is absent in the male if the testis is undescended

THE 'PHYSIOLOGICAL UMBILICAL HERNIA' OF THE FETUS:
115. Contains the fetal stomach
116. Occurs during the middle three months of gestation
117. Is reduced as the level of the diaphragm rises within the abdomen
118. Is inconstant
119. May persist until birth, as an abnormality

IN THE STOMACH:
120. A well-defined sphincter prevents reflux of contents into the oesophagus
121. Secretion is stimulated by the splanchnic nerves
122. The wall contains a layer of oblique muscle fibres
123. Oxyntic cells are found mainly in the antrum
124. The pyloric sphincter is ill-defined at birth

THE LIVER:
125. Is not essential to life
126. Has only one cell, the hepatocyte, which performs all of its essential functions
127. Lymphatics drain into nodes at portahepatis
128. Has no real capsule
129. Has sinusoidal spaces lined by reticuloendothelial cells

THE PANCREAS:
130. Develops as a single outgrowth from the embryonic foregut
131. Is strictly speaking entirely exocrine because all its secretory cells are derived from the endoderm
132. Empties its digestive enzymes into the pylorus
133. Its exocrine secretion is stimulated by anterior pituitary hormones
134. Lies in contact with both the duodenum and the spleen

THE PORTAL VEIN:
135. Is formed by union of superior and inferior mesenteric veins
136. Carries venous blood from the spleen
137. Runs in the lesser omentum with a branch of the coeliac artery
138. Notches the bare area of the liver
139. Obstruction may be manifested by dilated paraumbilical veins

THE GALL BLADDER:
140. Is lined by an epithelium which is modified for the primary purpose of absorption
141. Has a large number of mucous glands in its mucous membrane
142. Develops from the hepatic diverticulum
143. Has a well-developed muscularis externa
144. Receives motor innervation from the vagus nerve

HEPATOCYTES:
145. Have both exocrine and endocrine functions because they deliver products of their activity into both ducts and blood vessels
146. Form lobules which are the functional units of the liver
147. Are modified connective tissue cells of mesodermal origin
148. Are separated from the bile canaliculi by pavement endothelium
149. Can undergo mitosis to restore liver substance

THE SPLEEN:
150. Is a highly vascular organ
151. Is related to both the greater and lesser sacs of peritoneum
152. Normally lies in the axis of the left tenth rib
153. In the adult is normally palpable below the left rib margin
154. Cannot be palpated in the infant

THE DUODENUM:
155. Is retroperitoneal in its entire length
156. Is directly related to the right kidney
157. Lies anterior to the common bile duct
158. Lies posterior to the superior mesenteric vessels
159. Is an important site of portosystemic venous anastomosis

THE CRYPTS OF LIEBERKÜHN:
160. Are confined to the duodenum, jejunum and ileum
161. Are restricted to the mucous membrane
162. Secrete digestive enzymes
163. Contain enterochromaffin cells and Paneth cells
164. Receive the secretion of Brünner's glands

THE CAECUM:
165. Is non-distensible because of its relatively rigid wall
166. Contains liquid faeces and gas
167. Has well-developed taeniae coli
168. Occupies right upper abdomen in fetal life
169. Invariably occupies the right iliac fossa after birth

THE LARGE INTESTINE:
170. Has taenia coli in its muscularis externa
171. Contains fewer goblet cells than the small intestine
172. Has no villi
173. Has no plexus of Auerbach
174. Shows some stratification of lining epithelium distally

INTESTINAL VILLI:
175. Are confined to duodenum, jejunum and ileum
176. Are lined by columnar epithelial cells with microvillous free surfaces
177. Disappear when the intestine is distended
178. Contain abundant blood capillaries but no lymphatics
179. Each contains lamina propria and muscularis mucosae

THE ADRENAL CORTEX:
180. Is derived from neural crest ectoderm
181. Has a rich nerve supply
182. Is essential to life
183. Has three cellular zones recognizable by a characteristic arrangement of the cells in each zone
184. Is responsible for the production of the corticosteroids

THE ADRENAL MEDULLA:
185. Is derived from neural crest ectoderm
186. Has large polyhedral cholesterol-containing cells
187. Is essential to life
188. Stores adrenalin in an inactive form in its chromaffin cells
189. Venous blood drains into the portal system

THE ABDOMINAL AORTA:
190. Enters the abdomen behind the medial arcuate ligament
191. Normally gives origin directly to the splenic artery
192. Is crossed superficially by the left renal vein
193. Is related to many lymph nodes
194. Bifurcates at the level of the fifth lumbar vertebra

JEJUNUM

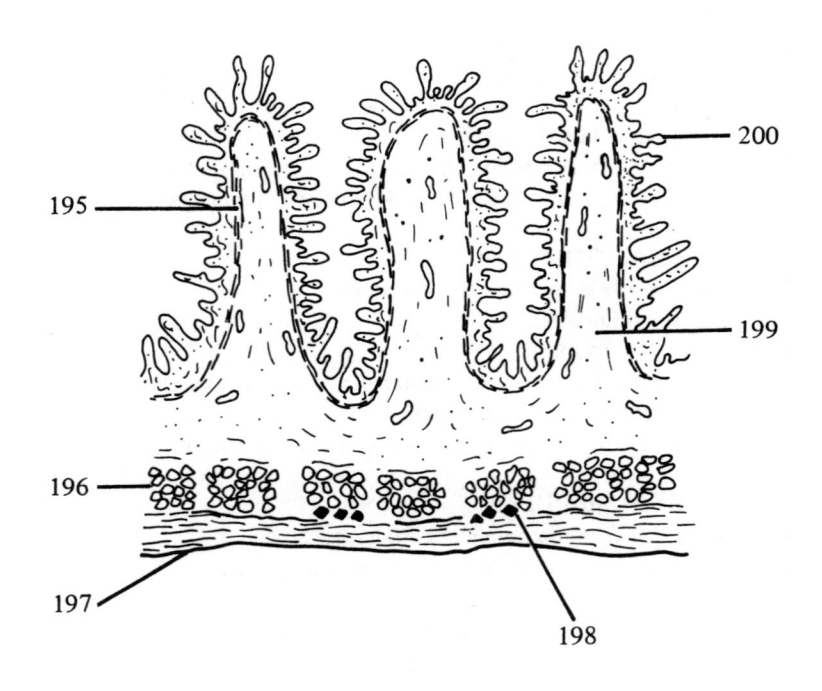

IDENTIFY THE NUMBERED STRUCTURES:

195 ___
196 ___
197 ___
198 ___
199 ___
200 ___

- A. Villus
- B. Submucous coat
- C. Auerbach's plexus
- D. Microvillus
- E. Serous coat
- F. Muscularis mucosae
- G. Meissner's plexus
- H. Lamina propria
- I. Crypt
- J. Circular muscle

POSTERIOR ABDOMINAL STRUCTURES

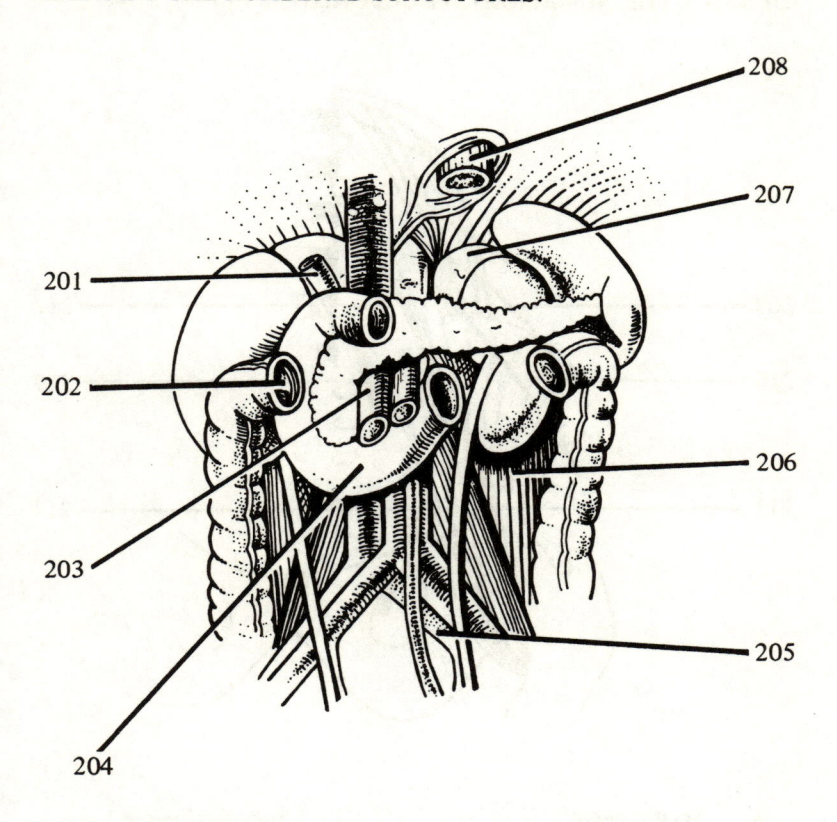

IDENTIFY THE NUMBERED STRUCTURES:

A.	Duodenum, third stage	I.	Duodenum, second stage
B.	Oesophagus	J.	Gastroduodenal vein
C.	Hepatic flexure of colon	K.	Suprarenal gland
D.	Common iliac vein	L.	Portal vein
E.	Splenic flexure of colon	M.	Splenic vein
F.	Quadratus lumborum	N.	Superior mesenteric artery
G.	Psoas major	O.	Inferior mesenteric artery
H.	Superior mesenteric vein	P.	Common iliac artery

81

NERVES ON POSTERIOR ABDOMINAL WALL

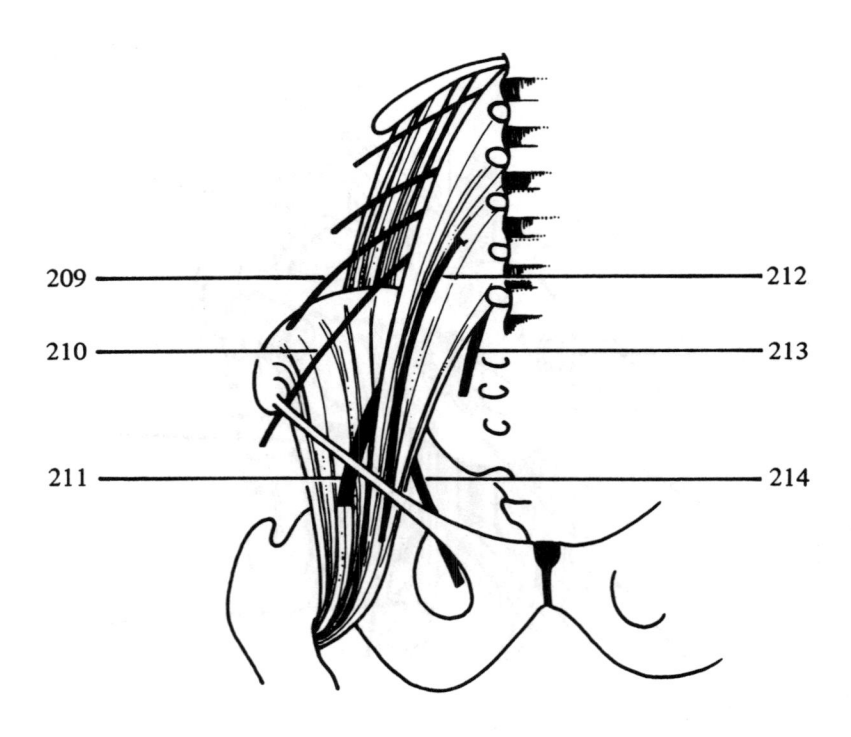

IDENTIFY THE NUMBERED STRUCTURES:

209 _____
210 _____
211 _____

212 _____
213 _____
214 _____

- A. Sciatic nerve
- B. Lumbosacral trunk
- C. Pudendal nerve
- D. Posterior cutaneous nerve of thigh
- E. Lateral cutaneous nerve of thigh
- F. Medial cutaneous nerve of thigh
- G. Subcostal nerve
- H. Femoral nerve
- I. Obturator nerve
- J. Genitofemoral nerve
- K. Iliohypogastric nerve
- L. Ilioinguinal nerve

ANSWERS

1.	E	36.	B	71.	D
2.	D	37.	B	72.	A
3.	D	38.	D	73.	C
4.	D	39.	B	74.	E
5.	D	40.	A	75.	B
6.	B	41.	D	76.	D
7.	E	42.	C	77.	E
8.	B	43.	C	78.	C
9.	A	44.	D	79.	A
10.	A	45.	D	80.	B
11.	A	46.	C	81.	A
12.	D	47.	C	82.	D
13.	B	48.	C	83.	C
14.	B	49.	D	84.	E
15.	A	50.	D	85.	E
16.	D	51.	A	86.	A
17.	C	52.	A	87.	B
18.	B	53.	B	88.	C
19.	E	54.	C	89.	D
20.	B	55.	C	90.	C
21.	D	56.	C	91.	A
22.	A	57.	C	92.	B
23.	A	58.	A	93.	E
24.	A	59.	C	94.	D
25.	B	60.	C	95.	C
26.	A	61.	D	96.	D
27.	B	62.	A	97.	E
28.	B	63.	C	98.	B
29.	C	64.	B	99.	A
30.	A	65.	D	100.	C
31.	A	66.	C	101.	D
32.	D	67.	A	102.	B
33.	A	68.	C	103.	E
34.	B	69.	A	104.	A
35.	B	70.	B	105.	T

106.	F	142.	T	178.	F
107.	T	143.	F	179.	F
108.	F	144.	T	180.	F
109.	F	145.	T	181.	F
110.	F	146.	F	182.	T
111.	F	147.	F	183.	T
112.	T	148.	F	184.	T
113.	F	149.	T	185.	T
114.	F	150.	T	186.	F
115.	F	151.	T	187.	F
116.	F	152.	T	188.	T
117.	F	153.	F	189.	F
118.	F	154.	F	190.	F
119.	T	155.	F	191.	F
120.	F	156.	T	192.	T
121.	F	157.	T	193.	T
122.	T	158.	T	194.	F
123.	F	159.	F	195.	F
124.	F	160.	T	196.	J
125.	F	161.	T	197.	E
126.	F	162.	T	198.	C
127.	T	163.	T	199.	B
128.	F	164.	T	200.	A
129.	T	165.	F	201.	L
130.	F	166.	T	202.	C
131.	F	167.	T	203.	H
132.	F	168.	F	204.	A
133.	F	169.	F	205.	D
134.	T	170.	T	206.	G
135.	F	171.	F	207.	K
136.	T	172.	T	208.	B
137.	T	173.	F	209.	L
138.	F	174.	F	210.	E
139.	T	175.	T	211.	H
140.	T	176.	T	212.	J
141.	F	177.	F	213.	B
				214.	I

SECTION V — PELVIS AND PERINEUM

QUESTIONS 1–17:

FOR EACH OF THE FOLLOWING MULTIPLE CHOICE QUESTIONS SELECT THE *ONE* MOST APPROPRIATE ANSWER:

1. THE PROSTATE:
    A. Is the size of an orange
    B. Surrounds the membranous urethra
    C. Is pierced by the ductus (vas) deferens
    D. Has no function in man
    E. None of the above

2. THE MIDDLE LOBE OF PROSTATE IS THE PART BETWEEN:
    A. Rectum and prostatic urethra
    B. Ejaculatory ducts and the rectum
    C. Ejaculatory ducts and the prostatic urethra
    D. Pubis and the prostatic urethra
    E. Pubis and rectum

3. THE PROSTATIC VENOUS PLEXUS COMMUNICATES WITH THE:
    A. Vesical venous plexus
    B. Vertebral venous plexus
    C. Deep dorsal vein of penis
    D. All of the above
    E. A and B only

4. THE DUCTUS (VAS) DEFERENS IS CONNECTED TO THE PROSTATIC URETHRA BY:
   A. The prostatic utricle
   B. Gartner's duct
   C. The ejaculatory duct
   D. The urachus
   E. None of the above

5. THE ANAL VALVES ARE AT THE LEVEL OF THE:
   A. Anorectal junction
   B. Anal margin
   C. White line of Hilton
   D. Rectal ampulla
   E. Pectinate line

6. THE UTERINE ARTERY:
   A. Usually arises from the internal iliac artery
   B. Anastomoses with the ovarian artery
   C. Crosses the ureter above the lateral vaginal fornix
   D. All of the above
   E. A and B only

7. THE CERVIX UTERI IS LINED BY TWO KINDS OF EPITHELIUM:
   A. Simple columnar and stratified squamous
   B. Ciliated columnar and stratified squamous
   C. Simple cubical and simple squamous
   D. Stratified squamous and simple squamous
   E. Transitional and columnar

8. THE SECRETORY PHASE OF THE MENSTRUAL CYCLE IS ASSOCIATED WITH:
   A. Increased vascularity of endometrium
   B. Saw-tooth appearance of uterine glands
   C. Maturation of the corpus luteum
   D. All of the above
   E. A and B only

9. THE INTERNAL OS OF THE UTERUS MARKS THE JUNCTION OF:
   A. Uterine tube and peritoneal cavity
   B. Uterine tube and fundus of uterus
   C. Fundus and body of uterus
   D. Body and cervix of uterus
   E. Cervix and vagina

10. THE LATERAL FORNIX OF THE VAGINA IS MOST CLOSELY RELATED TO THE:
    A. Urethra
    B. Ureter
    C. Middle rectal artery
    D. Uterine artery
    E. Round ligament of uterus

11. THE URORECTAL SEPTUM DIVIDES THE CLOACA INTO:
    A. Rectum and urogenital sinus
    B. Rectum and bladder
    C. Bladder and allantois
    D. Hindgut and bladder
    E. Urethra and anal canal

12. PERSISTENCE OF PORTION OF THE CLOACAL MEMBRANE MAY RESULT IN:
    A. An intact hymen
    B. An imperforate anus
    C. A bicornuate uterus
    D. All of the above
    E. A and B only

13. THE STEM CELLS LOCATED ON THE WALL OF THE SEMINIFEROUS TUBULE ARE:
    A. Sertoli cells
    B. Primary spermatocytes
    C. Secondary spermatocytes
    D. Spermatogonia
    E. Leydig cells

14. THE FOLLOWING STRUCTURES OCCUPY THE MALE SUPERFICIAL PERINEAL POUCH, *EXCEPT*:
    A. Corpora cavernosa
    B. Corpus spongiosum
    C. Posterior scrotal nerves and vessels
    D. Bulbo-urethral glands
    E. Bulb of penis

15. THE INFERIOR RECTAL NERVE SUPPLIES:
    A. External anal sphincter
    B. Internal anal sphincter
    C. Rectal mucous membrane
    D. Levator ani muscle
    E. Muscle of rectal ampulla

16. THE ANAL CANAL CONTAINS THE FOLLOWING EPITHELIUM:
    A. Stratified squamous
    B. Ciliated
    C. Simple squamous
    D. Stratified cuboidal
    E. None of the above

17. THE FOLLOWING CELLS MAKE CONTACT WITH THE BASEMENT MEMBRANE OF THE SEMINIFEROUS TUBULE:
    A. Spermatogonia and Sertoli cells
    B. Sertoli and Leydig cells
    C. Spermatogonia and Leydig cells
    D. Primary and secondary spermatocytes
    E. Spermatids

QUESTIONS 18–47:

THE SET OF LETTERED HEADINGS BELOW IS FOLLOWED BY A LIST OF NUMBERED WORDS OR PHRASES. FOR EACH NUMBERED WORD OR PHRASE SELECT THE CORRECT ANSWER UNDER:

    A.   If the item is associated with A only
    B.   If the item is associated with B only
    C.   If the item is associated with both A and B
    D.   If the item is associated with neither A nor B

    A.   Prostate
    B.   Bladder
    C.   Both
    D.   Neither

18. Lymph drainage to internal inguinal lymph nodes
19. Blood supply from inferior vesical artery
20. Pierced by ejaculatory ducts
21. Related to pelvic peritoneum
22. Contains the utricle

    A.   Related behind to rectum
    B.   Related in front to bladder
    C.   Both
    D.   Neither

23. Vagina
24. Seminal vesicle
25. Pouch of Douglas
26. Prostate
27. Cervix uteri

        A. Levator ani
        B. Sphincter ani internus
        C. Both
        D. Neither

28. Composed of striated muscle
29. Contains puborectalis
30. Contains sphincter urethrae
31. Motor supply from autonomic system
32. Attached to perineal body

        A. Endodermal epithelium
        B. Mesodermal epithelium
        C. Both
        D. Neither

33. Bladder
34. Urethra
35. Ureter
36. Prostatic glands
37. Rectum

        A. Passes through greater sciatic notch
        B. Passes through lesser sciatic notch
        C. Passes through both
        D. Passes through neither

38. Obturator externus
39. Obturator internus
40. Pudendal nerve
41. Sciatic nerve
42. Piriformis

A. Uterus
B. Ovary
C. Both
D. Neither

43. Lymph drainage to lumbar para-aortic nodes
44. Lymph drainage to internal iliac nodes
45. Nerve supply from nervi erigentes
46. Corpus luteum
47. Endometrium

QUESTIONS 48–54:

DIRECTIONS: In the following series of questions, one or more of the four items is/are correct. Answer A if 1, 2, 3 are correct; B if 1 and 3 are correct; C if 2 and 4 are correct; D if only 4 is correct; and E if all four are correct.

48. MALE ACCESSORY SEX GLAND(S):
    1. Seminal vesicles
    2. Prostate
    3. Bulbo-urethral
    4. Testis

49. SOURCE(S) OF ARTERIAL SUPPLY TO RECTUM:
    1. Inferior rectal
    2. Middle rectal
    3. Superior rectal
    4. External iliac

50. SITE(S) OF PORTO-SYSTEMIC ANASTOMOSIS:
    1. Rectosigmoid junction
    2. Prostatic plexus
    3. Cervix uteri
    4. Anorectal junction

51. THE PARAMESONEPHRIC DUCTS GIVE RISE TO THE ENTIRE EPITHELIUM OF:
    1. Uterine tubes
    2. Urethra
    3. Uterus
    4. Vagina

52. THE SUPERIOR HYPOGASTRIC PLEXUS (PRESACRAL NERVE) CONTAINS THE FOLLOWING NERVE FIBRES:
    1. Preganglionic parasympathetic
    2. Postganglionic sympathetic
    3. Postganglionic parasympathetic
    4. Visceral afferent

53. THE FEMALE PELVIS DIFFERS FROM THE MALE PELVIS IN THE FOLLOWING:
    1. Greater transverse diameter
    2. Greater anteroposterior diameter
    3. Shallower pelvic cavity
    4. Wider subpubic angle

54. THE LYMPHATIC VESSELS FROM THE TESTIS DRAIN INTO NODES:
    1. Below the inguinal ligament
    2. Along the internal iliac artery
    3. Along the external iliac artery
    4. Beside the aorta

QUESTIONS 55—69:

THE GROUP OF QUESTIONS BELOW CONSISTS OF NUMBERED HEADINGS, FOLLOWED BY A LIST OF LETTERED WORDS OR PHRASES. FOR EACH HEADING SELECT THE *ONE* WORD OR PHRASE WHICH IS MOST CLOSELY RELATED TO IT.
NOTE: EACH CHOICE MAY BE USED *MORE THAN ONCE.*

55.	Relationship of prostatic utricle to prostatic urethra	A.	Medial
56.	Relationship of ovary to obturator nerve	B.	Lateral
57.	Relationship of seminal vesicle to vas deferens	C.	Posterior
58.	Relationship of levator ani to ischio-rectal fossa	D.	Inferior
59.	Relationship of perineal membrane to prostate		

60.	Anal canal	A.	Internal iliac nodes
61.	Body of uterus	B.	Superficial inguinal nodes
62.	Testis	C.	Para-aortic nodes
63.	Bladder	D.	Deep inguinal nodes
64.	Vagina	E.	None of the above

65.	Ischial tuberosity	A.	Sacrotuberous ligament
66.	Ischial spine	B.	Sacrospinous ligament
67.	Body of pubis	C.	Pubovesical ligament
68.	Conjoint ramus	D.	Perineal membrane
69.	Pubic tubercle	E.	Inguinal ligament

QUESTIONS 70–119:

IN REPLY TO THE FOLLOWING QUESTIONS INDICATE WHETHER YOU THINK EACH STATEMENT IS *TRUE* OR *FALSE:*

THE PROSTATE GLAND:
70. Is situated in the perineum immediately below levator ani
71. Is surrounded by a plexus of veins
72. Has a stony hard consistency
73. Usually has a posterior median groove
74. Is palpable on rectal examination

THE BLADDER:
75. Is a pelvic organ in the infant
76. Has peritoneum on its anterior surface when full
77. Has an entirely smooth-walled interior
78. Its muscular walls are pierced obliquely by the ureters
79. The urachus is a 'safety valve' allowing overflow when the urethra is obstructed

THE RECTUM:
80. Contains shelf-like projections called rectal valves
81. Runs through the ischio-rectal fossa
82. Except for its upper third, is devoid of a peritoneal covering
83. Afferent nerves connect with sacral segments of spinal cord
84. Both Auerbach's and Meissner's plexuses are absent from this portion of the intestine

THE OVARIES:
85. Are covered by 'germinal epithelium' which gives origin to the germ cells
86. Contain about 400,000 primary follicles at birth
87. Undergo a tenfold increase in size at puberty
88. Develop a corpus luteum immediately prior to ovulation
89. Produce oestrogen in the follicular phase of the ovarian cycle

THE OVARIES:
90. Are attached to the posterior leaf of the broad ligaments of the uterus
91. Lie on the side wall of the pelvis in the angle between the internal and external iliac arteries
92. Receive their blood supply from the internal iliac arteries
93. Develop from the intermediate mesoderm in the embryo
94. Are normally palpable on abdominal examination

THE VAGINA:
95. Is lined by mucus-secreting columnar epithelium
96. Anterior wall is in contact with bladder and urethra
97. Posterior fornix is directly related to the recto-uterine pouch
98. Portion of levator ani forms a sling around its upper part
99. The vestibular glands (of Bartholin) open, each by a single duct, into its orifice

THE ANAL CANAL:
100. Differs structurally in male and female
101. Is separated from the prostate by peritoneum
102. Is highly sensitive in its lower half
103. Lymph drains to superficial inguinal nodes
104. Is surrounded by a voluntary sphincter

THE DUCT OF THE EPIDIDYMIS:
105. Connects the rete testis with the ductus (vas) deferens
106. Is lined by germinal epithelium
107. Occupies the body and tail of the epididymis
108. Contains smooth muscle in its wall
109. Develops from the paramesonephric duct

THE SEMINAL VESICLES:
110. Are surrounded by the corpora cavernosa
111. Store spermatozoa
112. Are lined by an absorptive type of epithelium
113. Have smooth muscle in their walls
114. Empty into the ejaculatory duct

THE PUDENDAL NERVE:
115. Arises from 2nd, 3rd and 4th sacral nerves
116. Crosses the sacrospinous ligament
117. Passes through the centre of the ischiorectal fossa
118. Usually gives origin to the inferior rectal nerve
119. Is sensory to the labia majora

RECTAL EXAMINATION

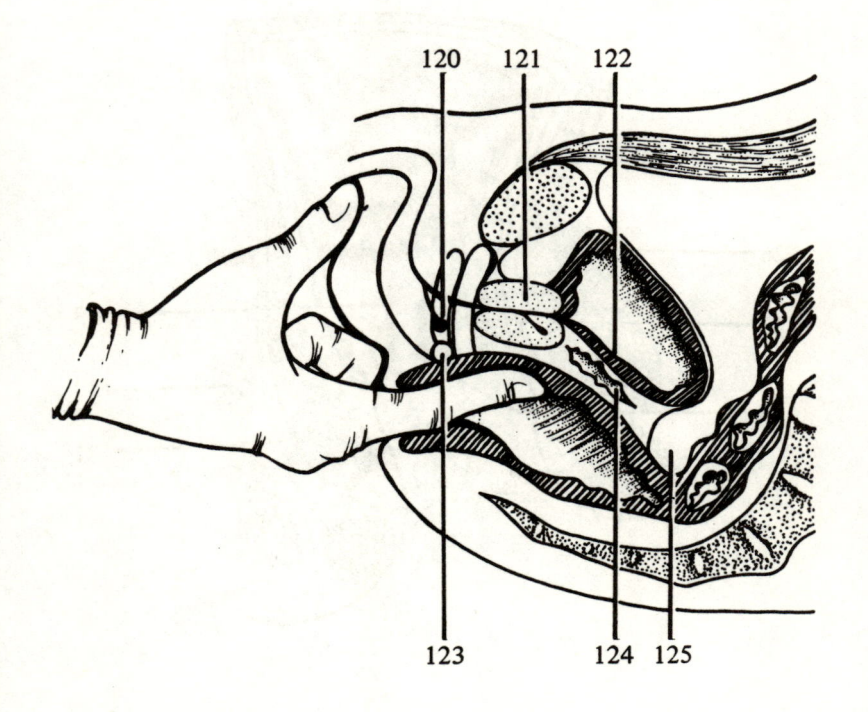

IDENTIFY THE NUMBERED STRUCTURES:

- A. Trigone
- B. Bulb of penis
- C. Prostate
- D. Vestibular gland
- E. Pouch of Douglas
- F. Ejaculatory duct
- G. Bulbourethral gland
- H. Seminal vesicle
- I. Pubic symphysis
- J. Perineal membrane
- K. Rectovesical pouch
- L. Perineal body

RECTUM AND ANAL CANAL

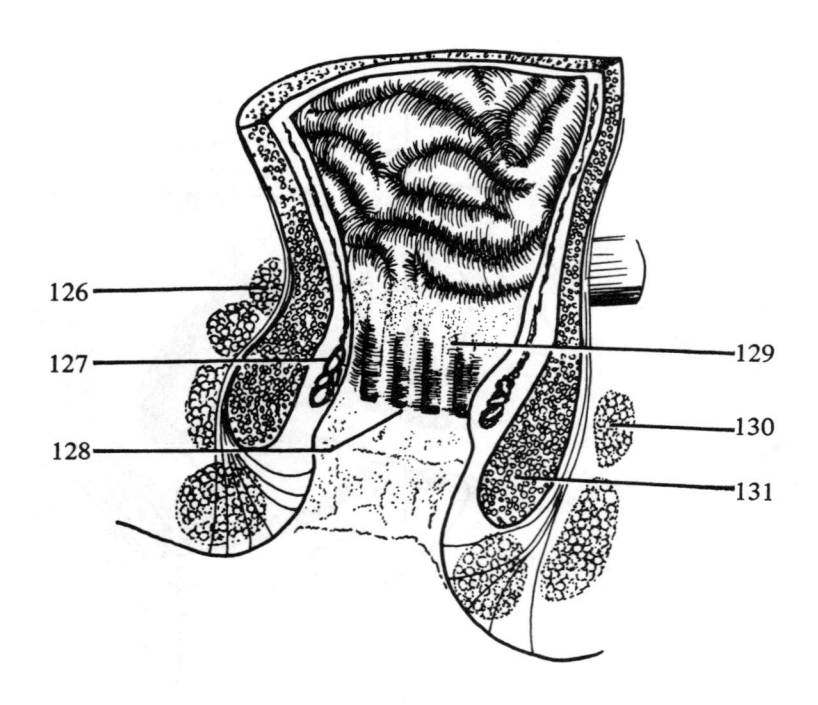

IDENTIFY THE NUMBERED STRUCTURES:

126 ⎯
127 ⎯
128 ⎯
129 ⎯
130 ⎯
131 ⎯

A.	Coccygeus	H.	Hilton's line
B.	Puborectalis	I.	Pecten
C.	Inferior rectal vein	J.	External sphincter, superficial part
D.	Anal valve		
E.	Longitudinal muscle	K.	Anal column
F.	Internal sphincter	L.	External sphincter, deep part
G.	Superior rectal vein		

ANSWERS

1.	E	36.	C	71.	T
2.	C	37.	A	72.	F
3.	D	38.	D	73.	T
4.	C	39.	B	74.	T
5.	E	40.	C	75.	F
6.	D	41.	A	76.	F
7.	A	42.	A	77.	F
8.	D	43.	B	78.	T
9.	D	44.	A	79.	F
10.	B	45.	A	80.	T
11.	A	46.	B	81.	F
12.	B	47.	A	82.	F
13.	D	48.	A	83.	T
14.	D	49.	A	84.	F
15.	A	50.	D	85.	F
16.	A	51.	B	86.	T
17.	A	52,	C	87.	F
18.	C	53.	E	88.	F
19.	C	54.	D	89.	T
20.	A	55.	C	90.	T
21.	B	56.	A	91.	T
22.	A	57.	B	92.	F
23.	C	58.	A	93.	T
24.	C	59.	D	94.	F
25.	A	60.	B	95.	F
26.	A	61.	A	96.	T
27.	C	62.	C	97.	T
28.	A	63.	A	98.	T
29.	A	64.	B	99.	T
30.	D	65.	A	100.	F
31.	B	66.	B	101.	F
32.	A	67.	C	102.	T
33.	C	68.	D	103.	T
34.	A	69.	E	104.	T
35.	B	70.	F	105.	T

106.	F	115.	T	124.	H
107.	T	116.	T	125.	K
108.	T	117.	F	126.	B
109.	F	118.	T	127.	G
110.	F	119.	T	128.	D
111.	F	120.	G	129.	K
112.	F	121.	C	130.	L
113.	T	122.	A	131.	F
114.	T	123.	L		

SECTION VI – HEAD AND NECK

QUESTIONS 1–54:

FOR EACH OF THE FOLLOWING MULTIPLE CHOICE QUESTIONS SELECT THE *ONE* MOST APPROPRIATE ANSWER:

1. THE CONTRACTING STERNOMASTOID:
   A. Rotates the head to the opposite side
   B. Tilts the head to the same side
   C. May assist respiration
   D. All of the above
   E. A and B only

2. THE COMMON CAROTID ARTERY ENDS AT THE LEVEL OF THE:
   A. Sternoclavicular joint
   B. Upper border of cricoid cartilage
   C. Upper border of thyroid cartilage
   D. Lower border of mandible
   E. Neck of mandible

3. THE FOLLOWING STRUCTURES OCCUPY THE CAROTID SHEATH, *EXCEPT*:
   A. Common carotid artery
   B. Internal carotid artery
   C. Vagus nerve
   D. Sympathetic trunk
   E. Internal jugular vein

4. SECTION OF THE INFERIOR (RECURRENT) LARYNGEAL NERVE WOULD PARALYSE THE INTRINSIC LARYNGEAL MUSCLES, *EXCEPT*:
   A. Cricothyroid
   B. Posterior cricoarytenoid
   C. Vocalis
   D. Lateral cricoarytenoid
   E. Interarytenoid

5. THE THYROID GLAND IS ENVELOPED IN:
   A. Investing fascia of the neck
   B. Prevertebral fascia
   C. Pretracheal fascia
   D. Superficial fascia
   E. None of the above

6. THE CRICOID CARTILAGE IS AT THE VERTEBRAL LEVEL OF:
   A. C2
   B. C4
   C. C6
   D. T1
   E. T3

7. ANTERIOR RELATION(S) OF SCALENUS ANTERIOR:
   A. Subclavian vein
   B. Phrenic nerve
   C. Brachial plexus
   D. Subclavian artery
   E. A and B

8. THE INTERMEDIATE TENDON OF THE OMOHYOID MUSCLE IS ANCHORED BY A FASCIAL SLING TO THE:
   A. Hyoid bone
   B. Thyroid cartilage
   C. Clavicle
   D. Sternum
   E. Scapula

9. THE CERVICAL PLEXUS IS FORMED OF THE VENTRAL RAMI OF:
   A. C1, C2, C3
   B. C1, C2
   C. C1, C2, C3, C4
   D. C3, C4
   E. C3, C4, C5

10. THE ROOTS OF THE BRACHIAL PLEXUS PASS BETWEEN:
    A. Scalenus anterior and sternomastoid
    B. Scalenus anterior and clavicle
    C. Scalenus anterior and scalenus medius
    D. Scalenus medius and scalenus posterior
    E. Scalenus posterior and levator scapulae

11. MUSCLE ATTACHED TO THE FIRST RIB BETWEEN SUBCLAVIAN VEIN AND ARTERY:
    A. Subclavius
    B. Scalenus medius
    C. Scalenus anterior
    D. Scalenus posterior
    E. Sternomastoid

12. THE LARGER BLOOD VESSELS OF THE SCALP RUN IN THE:
    A. Skin
    B. Subcutaneous tissue
    C. Epicranial aponeurosis
    D. Subaponeurotic tissue
    E. Pericranium

13. THE PLANE OF MOVEMENT OF THE SCALP IS BETWEEN:
    A. Skin and epicranial aponeurosis
    B. Epicranial aponeurosis (galea aponeurotica) and pericranium
    C. Skin and subcutaneous fat
    D. Pericranium and skull
    E. None of the above

14. THE MAIN SENSORY NERVE TO THE BACK OF THE HEAD:
    A. Greater auricular
    B. Greater occipital
    C. Posterior auricular
    D. Lesser occipital
    E. Third occipital

15. THE ZYGOMATIC ARCH GIVES ORIGIN TO:
    A. Masseter
    B. Temporalis
    C. Buccinator
    D. Platysma
    E. Medial pterygoid

16. MUSCLE(S) SUPPLIED BY THE FACIAL NERVE:
    A. Orbicularis oris
    B. Buccinator
    C. Orbicularis oculi
    D. All of the above
    E. A and B only

17. THE VERMILION BORDER OF THE LIP SHOWS:
    A. High dermal papillae bringing blood vessels close to the surface
    B. Abundant sebaceous glands
    C. Covering of simple squamous epithelium
    D. Large lymphoid nodules in submucosa
    E. All of the above

18. THE MAIN SENSORY NERVE TO THE UPPER LIP IS THE:
    A. Facial
    B. Infraorbital
    C. Buccal of mandibular
    D. External nasal
    E. Anterior superior alveolar

19. THE MAIN SENSORY NERVE TO THE LOWER LIP IS THE:
    A. Buccal
    B. Cervical of facial
    C. Mental
    D. Submandibular
    E. Inferior labial

20. THE MOST HORIZONTAL FIBRES OF TEMPORALIS MUSCLE ARE:
    A. Deep
    B. Superficial
    C. Posterior
    D. Middle
    E. Anterior

21. NERVE FROM WHICH THE PARASYMPATHETIC FIBRES FOR THE PAROTID GLAND TAKE ORIGIN:
    A. Vagus
    B. Facial
    C. Glossopharyngeal
    D. Trigeminal
    E. Accessory

22. A MUSCLE WHICH ELEVATES AND RETRACTS THE MANDIB:
    A. Masseter
    B. Temporalis
    C. Medial pterygoid
    D. Lateral pterygoid
    E. Digastric

23. A MUSCLE ATTACHED TO THE CORONOID PROCESS OF MANDIBLE:
    A. Buccinator
    B. Lateral pterygoid
    C. Medial pterygoid
    D. Temporalis
    E. Masseter

24. THE BUCCAL BRANCH OF THE MANDIBULAR NERVE SUPPLIES:
    A. Buccinator muscle
    B. Skin and mucous membrane of cheek
    C. Only skin of cheek
    D. Only mucous membrane of cheek
    E. Molar tooth pulp

25. THE FOLLOWING MUSCLE ASSISTS THE TONGUE IN KEEPING FOOD BETWEEN THE UPPER AND LOWER MOLAR TEETH DURING MASTICATION:
    A. Masseter
    B. Temporalis
    C. Orbicularis oris
    D. Buccinator
    E. Risorius

26. A STRUCTURE PASSING DEEP TO THE HYOGLOSSUS MUSCLE:
    A. Hypoglossal nerve
    B. Mylohyoid nerve
    C. Submandibular duct
    D. Lingual nerve
    E. None of the above

27. THE SUBMANDIBULAR TRIANGLE IS BORDED BY THE:
    A. Geniohyoid
    B. Mylohyoid
    C. Stylohyoid
    D. Digastric
    E. All of the above

28. THE PAROTID DUCT OPENS INTO THE ORAL CAVITY OPPOSITE THE CROWN OF THE:
    A. Upper first premolar tooth
    B. Upper second molar
    C. Lower first premolar
    D. Upper third molar
    E. Lower second molar

29. THE SUBLINGUAL PAPILLA:
   A. Is a swelling on the lingual nerve
   B. Is caused by the loop of the lingual artery
   C. Is a swelling caused by the sublingual gland
   D. Marks the opening of the submandibular duct
   E. Is a specialised taste bud

30. THE ORAL (BUCCOPHARYNGEAL) MEMBRANE SEPARATES:
   A. Amniotic sac from yolk sac
   B. Nasal cavity from oral cavity
   C. Proctodeum from hindgut
   D. Larynx from pharynx
   E. Stomodeum from foregut

31. THE MUSCLES OF THE TONGUE ARE DERIVED FROM:
   A. Occipital somites
   B. The mandibular arch
   C. The mandibular and hyoid arches
   D. Cervical somites
   E. None of the above

32. SENSORY NERVE(S) TO THE MUCOUS MEMBRANE LINING THE VESTIBULE OF THE MOUTH:
   A. Buccal branch of mandibular
   B. Infraorbital
   C. Mental
   D. All of the above
   E. A and B only

33. THE PERMANENT DENTITION IN EACH QUADRANT OF THE MOUTH COMPRISES:
   A. Two incisors, canine, two premolars, two molars
   B. One incisor, canine, two premolars, three molars
   C. Two incisors, canine, two premolars, three molars
   D. One incisor, canine, three premolars, three molars
   E. Two incisors, canine, one premolar, three molars

34. THE DECIDUOUS DENTITION IN EACH QUADRANT OF THE MOUTH COMPRISES:
    A. Two incisors, canine, two molars
    B. One incisor, canine, three molars
    C. One incisor, canine, one premolar, two molars
    D. Two incisors, canine, two premolars, two molars
    E. Two incisors, canine, three molars

35. THE FIRST PERMANENT TOOTH TO ERUPT IS THE:
    A. Lateral incisor
    B. Canine
    C. First premolar
    D. First molar
    E. Third molar

36. THE FIRST DECIDUOUS TOOTH TO ERUPT IS USUALLY THE:
    A. Central incisor
    B. Canine
    C. First premolar
    D. First molar
    E. Third molar

37. THE ODONTOBLAST IS RESPONSIBLE FOR THE PRODUCTION OF:
    A. Tooth enamel
    B. Dentin
    C. Cementum
    D. Cells of dentate nucleus
    E. None of the above

38. THE LINGUAL TONSIL IS CONFINED TO THE:
    A. Posterior one-third of the tongue
    B. Anterior one-third of the tongue
    C. Sulcus terminalis
    D. Ventral surface of the tongue
    E. None of the above

39. THE BONY PART OF THE NASAL SEPTUM CONTAINS:
    A. Perpendicular plate of ethmoid and vomer
    B. Perpendicular plate of palatine and vomer
    C. Perpendicular plate of palatine, vomer, and perpendicular plate of ethmoid
    D. Perpendicular plate of palatine, vomer, palatine crest of maxilla
    E. Perpendicular plate of vomer, palatine, rostrum of sphenoid

40. THE POSTERIOR EDGE OF THE SEPTAL CARTILAGE OF THE NOSE IS CONNECTED WITH THE:
    A. Middle concha
    B. Body of sphenoid
    C. Perpendicular plate of ethmoid and vomer
    D. Only the vomer
    E. Only the perpendicular plate of the ethmoid

41. THE ADULT MAXILLARY AIR SINUS ALWAYS LIES DIRECTLY ABOVE THE FOLLOWING TEETH:
    A. Incisors
    B. Incisors and canine
    C. Premolars
    D. Molars
    E. All of the above

42. A BRANCH OF THE MAXILLARY NERVE:
    A. Infraorbital
    B. Supraorbital
    C. Auriculotemporal
    D. External nasal
    E. Frontal

43. GLAND RECEIVING SECRETO-MOTOR FIBRES FROM THE PTERYGOPALATINE GANGLION:
    A. Lacrimal
    B. Submandibular
    C. Sublingual
    D. Parotid
    E. None of the above

44. THE FOLLOWING DURAL VENOUS SINUS OCCUPIES THE TENTORIUM CEREBELLI:
    A. Sphenoparietal
    B. Inferior petrosal
    C. Straight
    D. Inferior sagittal
    E. Occipital

45. STRUCTURE DIRECTLY ANTERIOR TO THE JUGULAR FORAMEN:
    A. Styloid process
    B. Occipital condyle
    C. Foramen magnum
    D. Sphenoidal spine
    E. Carotid canal

46. THE FOLLOWING BONY PARTS OSSIFY FROM CARTILAGINOUS PRECURSORS, *EXCEPT:*
    A. Greater wings of the sphenoid
    B. Lesser wings of the sphenoid
    C. Orbital plate of the frontal
    D. Petrous temporal
    E. Pre-occipital

47. THE FOLLOWING BELONG TO THE SPHENOID BONE, *EXCEPT:*
    A. Crista galli
    B. Anterior clinoid process
    C. Lateral pterygoid plate
    D. Sella turcica
    E. Optic foramen

48. IN THE NEUROHYPOPHYSIS, SECRETORY GRANULES ACCUMULATE IN:
    A. Pituicytes
    B. Nerve endings
    C. Intercellular spaces
    D. Capillary endothelium
    E. The lumen of sinusoids

49. THE SUPERIOR ORBITAL FISSURE IS BOUNDED BY THE:
    A. Maxilla and greater wing of sphenoid
    B. Maxilla and lesser wing of sphenoid
    C. Lesser wing of sphenoid and ethmoidal
    D. Lesser wing and greater wing of sphenoid
    E. Lesser wing of sphenoid and frontal

50. BONE(S) FORMING THE LATERAL MARGIN OF THE ORBIT:
    A. Zygomatic
    B. Frontal
    C. Zygomatic and frontal
    D. Zygomatic and sphenoidal
    E. Frontal and sphenoidal

51. CHARACTERISTIC(S) OF COMPLETE PARALYSIS OF THE SYMPATHETIC SUPPLY TO THE EYE:
    A. Ptosis
    B. Constriction of the pupil
    C. Inability to accommodate
    D. All of the above
    E. A and B only

52. OPENING OF THE MIDDLE EAR IN WHICH THE FOOTPLATE OF THE STAPES IS PLACED:
    A. Opening of pyramid
    B. Oval window (fenestra vestibuli)
    C. Aditus
    D. Tubal opening
    E. Round window (fenestra cochleae)

53. MAIN SENSORY NERVE TO THE PINNA OF THE EAR:
    A. Great auricular
    B. Greater occipital
    C. Zygomatico-facial
    D. Zygomatico-temporal
    E. Lesser occipital

54. THE ATLAS VERTEBRA IS UNIQUE IN HAVING:
    A. A foramen transversarium
    B. No body
    C. A bifid spine
    D. All of the above
    E. A and B only

QUESTIONS 55-94:

THE SET OF LETTERED HEADINGS BELOW IS FOLLOWED BY A LIST OF NUMBERED WORDS OR PHRASES. FOR EACH NUMBERED WORD OR PHRASE SELECT THE CORRECT ANSWER UNDER:

   A. If the item is associated with A only
   B. If the item is associated with B only
   C. If the item is associated with both A and B
   D. If the item is associated with neither A nor B

   A. Lateral pterygoid
   B. Medial pterygoid
   C. Both
   D. Neither

55. Attached to articular disk of temporomandibular joint
56. Attached to neck of mandible
57. Assists in opening the mouth
58. Assists in closing the mouth

A. Facial nerve
B. Mandibular nerve
C. Both
D. Neither

59. Motor to buccinator
60. Motor to medial pterygoid
61. Sensory to upper lip
62. Sensory to lower lip

A. Foramen ovale
B. Foramen spinosum
C. Both
D. Neither

63. Maxillary nerve
64. Mandibular nerve
65. Middle meningeal artery
66. Branches of maxillary artery

A. Hyoglossus
B. Genioglossus
C. Both
D. Neither

67. Hypoglossal nerve supply
68. Contact with submandibular gland
69. Superficial to lingual artery
70. Attached to mandible

A. Middle meatus of nose
B. Inferior meatus of nose
C. Both
D. Neither

71. Orifice of frontal air sinus
72. Orifice of nasolacrimal duct
73. Orifice of sphenoidal air sinus
74. Lined by olfactory epithelium
75. Origin from stomodaeum of embryo

A. Superior petrosal sinus
B. Sigmoid sinus
C. Both
D. Neither

76. Is an air sinus
77. Direct communication with cavernous sinus
78. Lodged in groove on temporal bone
79. Posterior relation of tympanic cavity

A. Superior rectus muscle of eye
B. Inferior oblique muscle of eye
C. Both
D. Neither

80. Oculomotor nerve supply
81. Looking upwards
82. Looking downwards
83. Looking downwards and outwards
84. Muscle spindles are present

A. Ophthalmic nerve
B. Facial nerve
C. Both
D. Neither

85. Light reflex
86. Accommodation reflex
87. Corneal reflex
88. Consensual reflex
89. Jaw jerk

A. Tympanic cavity
B. Epitympanic recess
C. Both
D. Neither

90. Related to facial nerve
91. Leads directly into mastoid antrum
92. Leads directly into Eustachian tube
93. Related directly to tympanic membrane
94. Derived from ectoderm of first pharyngeal groove (cleft)

QUESTIONS 95—113:

DIRECTIONS: In the following series of questions, one or more of the four items is/are correct. Answer A if 1, 2, 3 are correct; B if 1 and 3 are correct; C if 2 and 4 are correct; D if only 4 is correct; and E if all four are correct.

95. HAEMOSTATS PLACED ON THE INFERIOR THYROID ARTERY MAY ENDANGER:
    1. The inferior (recurrent) laryngeal nerve
    2. The external laryngeal nerve
    3. The cervical sympathetic chain
    4. The phrenic nerve

96. REMNANT(S) OF EMBRYONIC THYROID OUTGROWTH FROM THE PHARYNX:
    1. Branchial cyst
    2. Laryngocoele
    3. Cystic hygroma
    4. Pyramidal lobe

97. THE VOCAL FOLDS ARE ABDUCTED BY:
    1. Thyro-arytenoids
    2. Lateral crico-arytenoids
    3. Inter-arytenoids
    4. Posterior crico-arytenoids

98. THE VOCAL FOLDS ARE ATTACHED TO THE:
    1. Hyoid bone
    2. Thyroid cartilage
    3. Cricoid cartilage
    4. Arytenoid cartilages

99. THE CRICOPHARYNGEUS MUSCLE IS:
    1. Part of the middle constrictor of the pharynx
    2. Innervated by inferior (recurrent) laryngeal nerve
    3. Composed of smooth muscle fibres
    4. Horizontally disposed

100. STRATIFIED SQUAMOUS EPITHELIUM LINES THE:
    1. Floor of the mouth
    2. Laryngeal pharynx
    3. Oropharynx
    4. Nasopharynx

101. THE SENSORY SUPPLY OF THE PULPS OF THE UPPER TEETH IS BY:
    1. Posterior superior alveolar nerve
    2. Middle superior alveolar nerve
    3. Anterior superior alveolar nerve
    4. Infraorbital nerve

102. THE SOFT PALATE:
    1. Develops from paired embryonic maxillary processes
    2. Is elevated mainly by tensor palati
    3. May be occupied by portion of the tonsil
    4. Contains the horizontal plate of the palatine bone

103. THE EPITHELIUM OF THE POSTERIOR THIRD OF THE TONGUE:
    1. Is elevated by submucosal lymph follicles
    2. Develops from the embryonic endoderm
    3. Gives lymphatic drainage to the jugulo-omohyoid nodes
    4. Receives taste fibres from the chorda tympani nerve

104. THE FOREGUT OF THE EMBRYO GIVES RISE TO THE:
    1. Anterior lobe of hypophysis
    2. Thyroid gland
    3. Inner ear
    4. Lower respiratory tract

105. THE STRAIGHT SINUS:
    1. Occupies the line of attachment of falx cerebri to vault of skull
    2. Is formed by union of the great cerebral vein and inferior sagittal sinus
    3. Is commonly absent
    4. Enters the confluence of sinuses

106. THE UPPER EYELID:
    1. Contains modified sebaceous glands
    2. Is strengthened by the tarsus
    3. Is raised by contraction of levator palpebrae superioris
    4. Is lowered by contraction of orbicularis oculi

107. THE CORNEA:
    1. Is mainly composed of collagen bundles
    2. Contains no blood vessels
    3. Contains no lymphatics
    4. Contains only free nerve endings

108. THE LACRIMAL SAC:
    1. Lies against the upper lateral wall of the orbit
    2. Is emptied by blinking
    3. Secretes the tears
    4. Opens into the nasolacrimal duct

109. ACCOMMODATION OF THE EYE IS ASSOCIATED WITH CONTRACTION OF THE:
    1. Dilator pupillae
    2. Sphincter pupillae
    3. Orbicularis oculi
    4. Ciliary muscle

110. THE ABDUCENT NERVE INNERVATES THE:
  1. Superior rectus
  2. Medial rectus
  3. Inferior rectus
  4. Lateral rectus

111. THE NERVE FIBRE LAYER OF THE RETINA:
  1. Is next to the pigment layer
  2. Is opaque to light
  3. Synapses with bipolar cells
  4. Is composed of the central processes of ganglionic cells

112. THE DUCT OF THE COCHLEA:
  1. Contains endolymph
  2. Is separated from scala vestibuli by the vestibular (spiral) membrane
  3. Has the basilar membrane (lamina) in its floor
  4. Is separated from scala tympani by the tympanic membrane

113. THE EXTERNAL ACOUSTIC MEATUS:
  1. Has cartilaginous and bony parts
  2. Is relatively short in infancy
  3. Is lined by skin
  4. Develops from first pharyngeal groove (cleft)

QUESTIONS 114–140:

THE GROUP OF QUESTIONS BELOW CONSISTS OF NUMBERED HEADINGS, FOLLOWED BY A LIST OF LETTERED WORDS OR PHRASES. FOR EACH HEADING SELECT THE *ONE* WORD OR PHRASE WHICH IS MOST CLOSELY RELATED TO IT.
NOTE: EACH CHOICE MAY BE USED *ONCE ONLY.*

114. Submental lymph nodes
115. Submandibular lymph nodes
116. Preauricular lymph nodes
117. Mastoid lymph nodes

A. Tip of tongue
B. Lower molar teeth
C. Outer canthus of eye
D. Posterior third of tongue
E. None of the above

118. Soft palate
119. Anterior part of tongue
120. External acoustic meatus
121. Tympanic cavity
122. Orbicularis oris

A. Maxillary process
B. Mandibular arch
C. First pharyngeal groove
D. First pharyngeal pouch
E. Hyoid arch

123. Internal carotid artery
124. External carotid artery
125. Vertebral artery
126. Maxillary artery
127. Middle meningeal artery

A. Pterygomaxillary fissure
B. Foramen magnum
C. Carotid foramen
D. Foramen spinosum
E. None of the above

128. Maxillary nerve
129. Nervus spinosus
130. Mandibular nerve
131. Facial nerve
132. Glossopharyngeal nerve

A. Foramen ovale
B. Foramen spinosum
C. Stylomastoid foramen
D. Jugular foramen
E. Foramen rotundum

133. Internal acoustic meatus
134. Stylomastoid foramen
135. Incisive canal
136. Hypoglossal canal

A. Nasopalatine nerve
B. Facial nerve
C. Hypoglossal nerve
D. Middle meningeal vein
E. None of the above

137. First branchial arch
138. Second branchial arch
139. Third branchial arch
140. Fourth branchial arch

A. Mandible
B. Greater cornu of hyoid bone
C. Styloid process
D. Thyroid cartilage
E. Clavicle

QUESTIONS 141–290:

IN REPLY TO THE FOLLOWING QUESTIONS INDICATE WHETHER YOU THINK EACH STATEMENT IS *TRUE* OR *FALSE:*

THE EXTERNAL JUGULAR VEIN:
141. Is formed by the junction of the superficial temporal and maxillary veins
142. Crosses the sternomastoid to enter the posterior triangle
143. Because it lies beneath the deep fascia, is never visible in the living
144. Usually terminates in the subclavian vein
145. Because it lacks competent valves, it is an important indicator of pressure changes in the right atrium

THE POSTERIOR TRIANGLE OF THE NECK:
146. Is bordered by sternomastoid, mandible and trapezius
147. Has the eleventh nerve in its floor
148. Is confined to the posterior aspect of the neck
149. Contains the trunks of the brachial plexus
150. Contains numerous lymph nodes

THE SPINAL ACCESSORY NERVE:
151. Arises entirely from the spinal cord
152. Enters the skull via the foramen magnum
153. Leaves the skull via the jugular foramen
154. Contributes to the pharyngeal plexus of nerves
155. Injury in the posterior triangle results in wasting of the trapezius on the same side

THE THYROID GLAND:
156. Is closely related to inferior (recurrent) laryngeal nerve
157. Moves on swallowing
158. Isthmus is at level of thyroid cartilage
159. Is included in pretracheal fascia together with the parathyroids
160. Accessory thyroid tissue is sometimes present in region of internal jugular vein

THE THYROID GLAND:
161. Develops as an endodermal downgrowth from the floor of the primitive mouth
162. Has rich capillary plexuses closely related to its follicular cells
163. Resting follicles contain a small store of colloid surrounded by tall columnar epithelium
164. Colloid vacuoles in the epithelial cells are concerned with the synthesis of thyroglobulin
165. Overactivity is accompanied by hypertrophy of the Golgi apparatus and endoplasmic reticulum in the epithelial cells

THE PARATHYROID GLANDS:
166. Lie in close relationship to the thyroid gland on its anterior surface
167. Contain chief (principal) cells which secrete the hormone and contain an abundant rough surfaced endoplasmic reticulum
168. Have an elaborate system of excretory ducts
169. Secrete a hormone which causes an elevation in the level of blood calcium
170. Are controlled by anterior pituitary gland hormones

IN THE LARYNX:
171. Epithelium above the vocal folds is mainly stratified squamous
172. Sensory innervation is entirely from superior laryngeal nerves
173. Motor supply to intrinsic muscles is entirely from inferior (recurrent) laryngeal nerves
174. Posterior cricoarytenoid muscle abducts the vocal cords
175. Respiration in infancy is aided by its relatively wide lumen

THE OPHTHALMIC NERVE:
176. Enters the orbit through the optic foramen
177. Carries corneal sensation
178. Is motor to levator or palpebrae superioris
179. Is sensory to the upper teeth
180. When injured the light reflex is abolished on the same side

THE MAXILLARY NERVE:
181. Arises from the trigeminal ganglion
182. Occupies the lateral wall of the cavernous sinus
183. Has the otic ganglion attached to it
184. Supplies the skin of the bridge of the nose
185. Is entirely sensory

THE FACIAL NERVE:
186. Leaves the skull through foramen spinosum
187. Supplies sensory fibres to the conjunctiva
188. Is motor to the buccinator
189. Carries secreto-motor fibres for the parotid gland
190. Provides one limb of the corneal reflex

THE HYPOGLOSSAL NERVE:
191. Supplies sensation to the anterior part of the tongue
192. Carries fibres from the second and third cervical nerves
193. Passes between the internal and external carotid arteries
194. Is motor to both intrinsic and extrinsic muscles of the tongue
195. Following paralysis, the tongue on protrusion deviates towards the affected side

IN THE TEMPORO-MANDIBULAR JOINT:
196. Contact is between mandible and tympanic plate
197. Articular surface includes the eminentia articularis
198. Gliding movements take place in the upper compartment
199. Rotatory movement takes place in the lower compartment
200. Axis of joint rotation passes through the head of the mandible

THE PAROTID GLAND:
201. Is wedged between trapezius and mandible
202. Is covered by investing layer of cervical fascia
203. Contains the facial artery
204. Has lymph nodes buried in its substance
205. Has one duct which pierces buccinator to enter the mouth

THE SUBMANDIBULAR SALIVARY GLAND:
206. Has two lobes
207. Discharges into the mouth via the sublingual papilla
208. Lies entirely deep to the mylohyoid
209. Receives secreto-motor fibres from the submandibular ganglion
210. Has only one duct

THE SUBLINGUAL GLAND:
211. Occupies the sublingual fold in floor of mouth
212. Lies between mylohyoid and digastric muscles
213. Has only one large duct
214. Secreto-motor nerves travel via chorda tympani and lingual nerves
215. Lies on the submandibular duct

THE LINGUAL NERVE:
216. Leaves the skull through the pterygomaxillary fissure
217. Carries fibres from the first cervical spinal nerve
218. Is closely related to the mandible behind the third molar tooth
219. Carries secreto-motor fibres to the submandibular salivary gland
220. Carries fibres for both taste and common sensation from the anterior tongue

THE TONGUE:
221. Contains smooth muscle fibres running in three directions
222. Contains many taste buds on its ventral (inferior) surface
223. Is divisible into an anterior, papillary two-thirds and a posterior, lymphoid one-third
224. Contains no glands
225. Is anchored only by the mandible

TOOTH ENAMEL:
226. Is the hardest tissue in the body
227. Is replaced, when worn or decayed, by action of the ameloblasts
228. Is richly innervated
229. Is the only part of the tooth which is derived from ectoderm
230. Is impermeable

DENTIN:
231. Is of mesodermal origin
232. Cannot be replaced once it is destroyed
233. Is insensitive
234. Forms the bulk of the tooth
235. Contains its formative cells, the odontoblasts, evenly dispersed throughout its intercellular substance

CEMENTUM:
236. Is derived from epithelial cells called cementoblasts
237. Forms the outer layer of the root of the tooth
238. Makes direct contact with the enamel
239. Is acellular
240. Gives attachment to the periodontal ligament

THE HARD PALATE:
241. Is lined on its oral surface by columnar epithelium with goblet cells
242. Is attached to the nasal septum
243. Is concerned with mastication but not with speech
244. Has a mucous membrane which is loosely attached to the underlying bone to allow free movement
245. Contains no glands

THE SOFT PALATE:
246. Has the same function as the hard palate in mastication of food
247. Has keratinized epithelium on its oral surface
248. Contains a core of hyaline cartilage
249. Forms a boundary of the nasopharynx
250. Plays an important part in swallowing

THE PHARYNX:
251. Is lined throughout by stratified squamous epithelium
252. Has longitudinal and circular muscle layers
253. Lamina propria contains lymphoid tissue
254. Contains glands which open into the tonsillar crypts and help wash out debris
255. Is derived from the embryonic foregut

THE MAXILLARY AIR SINUS:
256. Is rudimentary at birth
257. Is lined by columnar ciliated epithelium
258. Contains the infra-orbital nerve in its roof
259. Opens into the middle meatus of the nose
260. Is innervated by the superior alveolar nerves

IN THE PITUITARY GLAND:
261. The adenohypophysis comprises pars anterior, pars intermedia and pars tuberalis
262. Capillary blood passes from the adenohypophysis upwards to the median eminence
263. The chromophobe cells of the adenohypophysis are non-secretory
264. Tumours of the basophil cells cause excessive production of growth hormone
265. The capillary plexus in the adenohypophysis is accompanied by axons containing neurosecretory material

THE MIDDLE MENINGEAL ARTERY:
266. Is a branch of the internal carotid
267. Enters the skull through the foramen spinosum
268. Gives a branch to the circle of Willis
269. Divides in the region of the pterion
270. Haemorrhage from the artery is usually subdural

IN THE EYE:
271. Cornea is normally devoid of blood vessels
272. Lens is situated between anterior and posterior chambers
273. Iris is constricted by stimulation of the oculomotor nerve
274. Vitreous humour drains into canal of Schlemm
275. Maximal light sensitivity is at optic disc

THE OPHTHALMIC ARTERY:
276. Arises from the third part of the maxillary artery
277. Enters orbit through the optic foramen
278. Carries nerves which constrict the pupil
279. Anastomoses with branches of the facial artery
280. Occlusion of its retinal branch causes blindness

IN THE DEVELOPING EYE:
281. The lens arises from the surface ectoderm
282. The optic stalk grows out from the telencephalon
283. The fetal fissure is without developmental significance in man
284. Blood vessels pass through the posterior chamber to reach the back of the lens
285. Lacrimal secretion does not commence until three weeks after birth

THE MIDDLE EAR:
286. Is essential for the conduction of sound
287. Receives sensory fibres from the eighth cranial nerve
288. Is related posteriorly to the sigmoid sinus
289. Is almost fully developed at birth
290. Cavity communicates with mastoid air cells

HORIZONTAL SECTION OF NECK

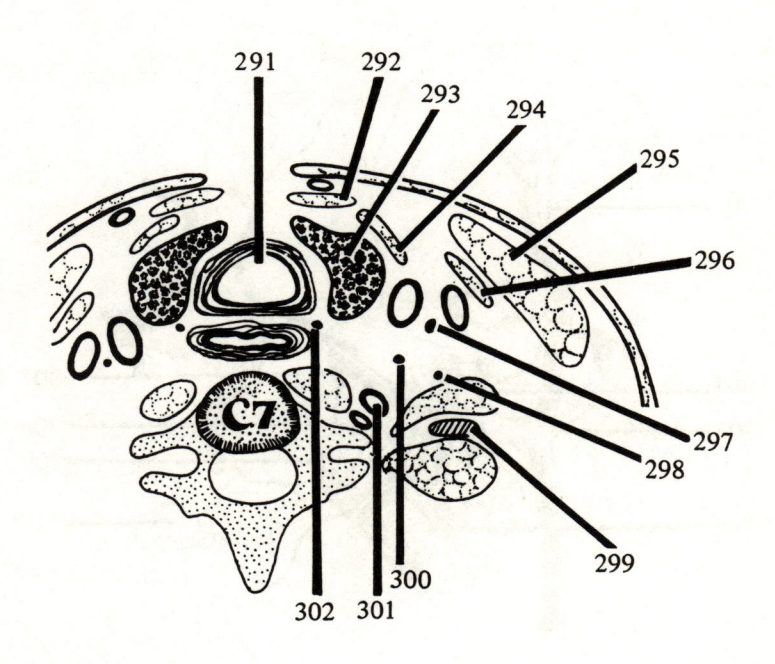

IDENTIFY THE NUMBERED STRUCTURES:

A.	Scalenus medius	I.	Thoracic duct
B.	Phrenic nerve	J.	Trachea
C.	Thyroid gland	K.	Omohyoid
D.	Vagus nerve	L.	Sternothyroid
E.	Scalenus anterior	M.	Sternohyoid
F.	Inferior laryngeal nerve	N.	Sternomastoid
G.	Superior laryngeal nerve	O.	Brachial plexus
H.	Vertebral vein	P.	Sympathetic trunk

PTERYGOID REGION

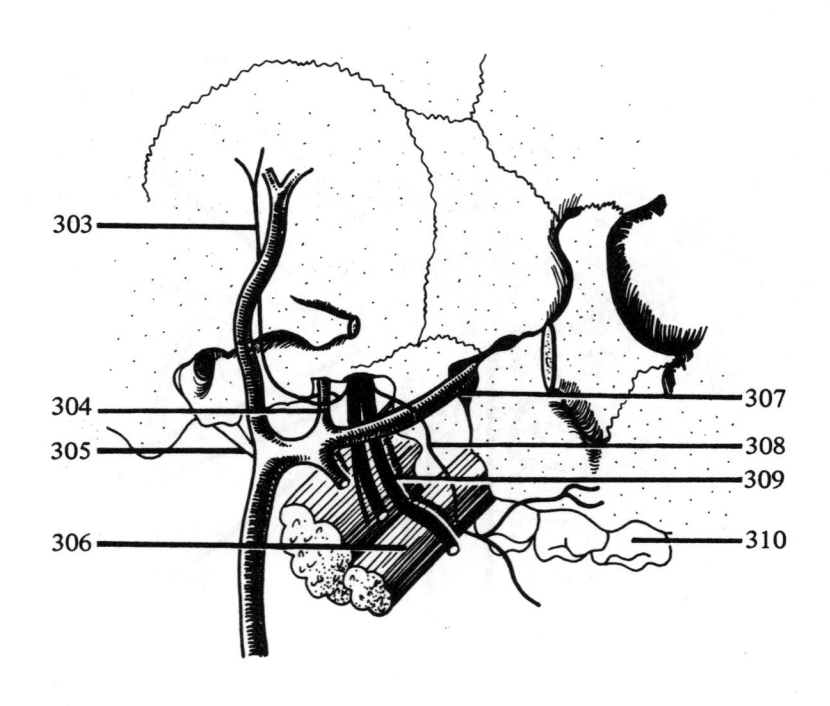

IDENTIFY THE NUMBERED STRUCTURES:

- A. Styloid process
- B. Medial pterygoid
- C. Lateral pterygoid
- D. Foramen ovale
- E. Deep temporal nerve
- F. Second premolar tooth
- G. First molar tooth
- H. Auriculotemporal nerve
- I. Lingual nerve
- J. Inferior alveolar nerve
- K. Middle meningeal artery
- L. External carotid artery
- M. Buccal nerve
- N. Pterygomaxillary fissure

CAVERNOUS SINUS

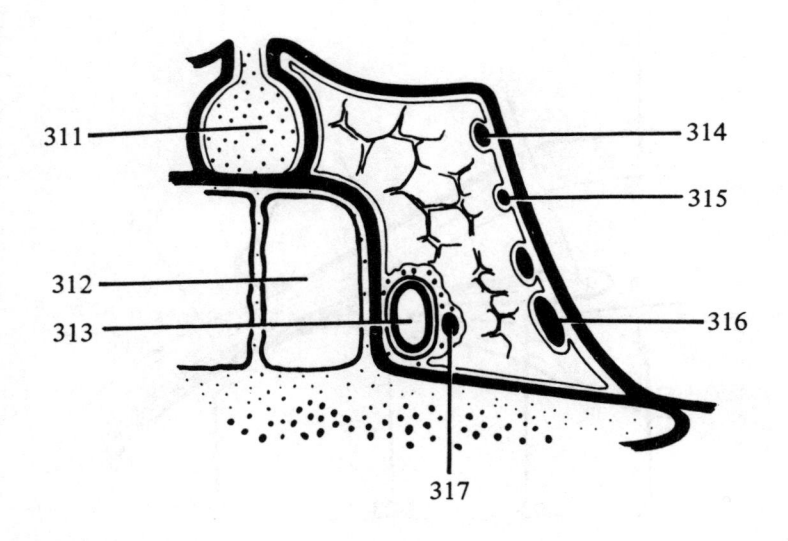

IDENTIFY THE NUMBERED STRUCTURES:

A.	Ophthalmic artery	G.	Internal carotid artery
B.	Sphenoidal sinus	H.	Oculomotor nerve
C.	Ethmoid sinus	I.	Trochlear nerve
D.	Abducent nerve	J.	Mandibular nerve
E.	Nasal cavity	K.	Hypophysis
F.	Maxillary nerve	L.	Middle meningeal vein

ANTERIOR PART OF EYE

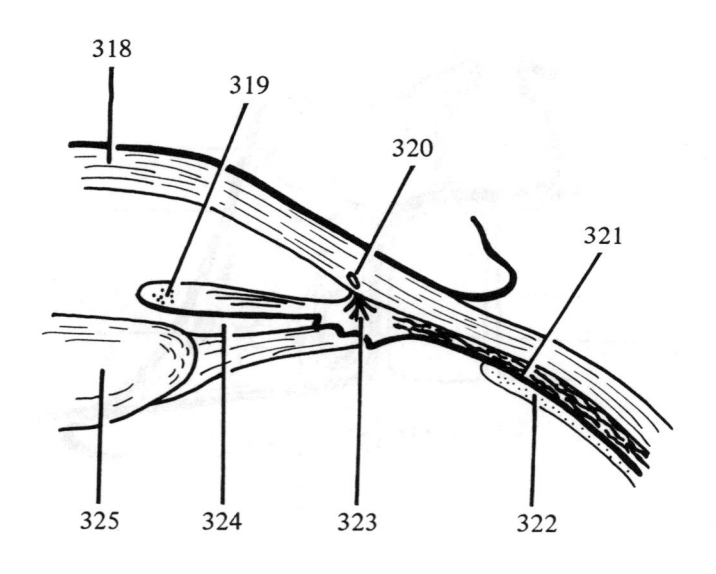

IDENTIFY THE NUMBERED STRUCTURES:

318
319
320
321
325 324 323 322

A.	Choroid	H.	Ciliaris
B.	Cornea	I.	Ophthalmic vein
C.	Sclera	J.	Anterior chamber
D.	Sphincter pupillae	K.	Sinus venosus sclerae
E.	Suspensory ligament	L.	Dilator pupillae
F.	Posterior chamber	M.	Vitreous
G.	Lens	N.	Retina

ANSWERS

1.	D	36.	A	71.	A
2.	C	37.	B	72.	B
3.	D	38.	A	73.	D
4.	A	39.	A	74.	D
5.	C	40.	C	75.	C
6.	C	41.	D	76.	D
7.	E	42.	A	77.	A
8.	C	43.	A	78.	C
9.	C	44.	C	79.	B
10.	C	45.	E	80.	C
11.	C	46.	C	81.	C
12.	B	47.	A	82.	D
13.	B.	48.	B	83.	D
14.	B	49.	D	84.	C
15.	A	50.	C	85.	D
16.	D	51.	E	86.	D
17.	A	52.	B	87.	C
18.	B	53.	A	88.	D
19.	C	54.	B	89.	D
20.	C	55.	A	90.	C
21.	C	56.	A	91.	B
22.	B	57.	A	92.	A
23.	D	58.	B	93.	A
24.	B	59.	A	94.	D
25.	D	60.	B	95.	B
26.	E	61.	D	96.	D
27.	D	62.	B	97.	D
28.	B	63.	D	98.	C
29.	D	64.	A	99.	C
30.	E	65.	B	100.	A
31.	A	66.	C	101.	A
32.	D	67.	C	102.	B
33.	C	68.	A	103.	A
34.	A	69.	A	104.	C
35.	D	70.	B	105.	C

106.	E	145.	T	184.	F
107.	E	146.	F	185.	T
108.	C	147.	T	186.	F
109.	C	148.	F	187.	F
110.	D	149.	T	188.	T
111.	D	150.	T	189.	F
112.	A	151.	T	190.	T
113.	E	152.	T	191.	F
114.	A	153.	T	192.	F
115.	B	154.	F	193.	F
116.	C	155.	T	194.	T
117.	E	156.	T	195.	T
118.	A	157.	T	196.	F
119.	B	158.	F	197.	T
120.	C	159.	T	198.	T
121.	D	160.	F	199.	T
122.	E	161.	T	200.	F
123.	C	162.	T	201.	F
124.	E	163.	F	202.	T
125.	B	164.	F	203.	F
126.	A	165.	T	204.	T
127.	D	166.	F	205.	T
128.	E	167.	T	206.	T
129.	B	168.	F	207.	T
130.	A	169.	T	208.	F
131.	C	170.	F	209.	T
132.	D	171.	T	210.	T
133.	B	172.	F	211.	T
134.	B	173.	F	212.	F
135.	A	174.	T	213.	F
136.	C	175.	F	214.	T
137.	A	176.	F	215.	T
138.	C	177.	T	216.	F
139.	B	178.	F	217.	F
140.	D	179.	F	218.	T
141.	F	180.	F	219.	T
142.	T	181.	T	220.	T
143.	F	182.	T	221.	F
144.	T	183.	F	222.	F

223.	T	257.	T	291.	J
224.	F	258.	T	292.	M
225.	F	259.	T	293.	C
226.	T	260.	T	294.	L
227.	F	261.	T	295.	N
228.	F	262.	F	296.	K
229.	T	263.	T	297.	D
230.	F	264.	F	298.	B
231.	T	265.	F	299.	O
232.	F	266.	F	300.	P
233.	F	267.	T	301.	H
234.	T	268.	F	302.	F
235.	F	269.	T	303.	H
236.	T	270.	F	304.	K
237.	T	271.	T	305.	A
238.	T	272.	F	306.	B
239.	F	273.	T	307.	N
240.	T	274.	F	308.	M
241.	F	275.	F	309.	I
242.	T	276.	F	310.	G
243.	F	277.	T	311.	K
244.	F	278.	F	312.	B
245.	F	279.	T	313.	G
246.	F	280.	T	314.	H
247.	T	281.	T	315.	I
248.	F	282.	F	316.	F
249.	T	283.	F	317.	D
250.	T	284.	T	318.	B
251.	F	285.	F	319.	D
252.	F	286.	F	320.	K
253.	T	287.	F	321.	A
254.	F	288.	T	322.	N
255.	T	289.	T	323.	H
256.	T	290.	T	324.	F
				325.	G

SECTION VII – NERVOUS SYSTEM

QUESTIONS 1–76:

FOR EACH OF THE FOLLOWING MULTIPLE CHOICE QUESTIONS SELECT THE *ONE* MOST APPROPRIATE ANSWER:

1. IMMEDIATELY EXTERNAL TO THE MYELIN SHEATH OF A NERVE FIBRE IS:
    A. Endoneurium
    B. Schwann cytoplasm
    C. Neurokeratin
    D. Basement membrane
    E. Axolemma

2. IN THE "MOTOR NERVE TRUNK" ENTERING A MUSCLE, SENSORY FIBRES MAKE UP THE FOLLOWING PROPORTION:
    A. None
    B. 20%
    C. 40%
    D. 75%
    E. 90%

3. THE FOLLOWING NERVE ENDINGS ARE FOUND IN THE SKIN:
    A. Meissner's corpuscles
    B. Bulbous corpuscles
    C. Pacinian corpuscles
    D. All of the above
    E. A and B only

4. PACINIAN CORPUSCLES ARE NUMEROUS IN:
   A. The dermis of the skin
   B. Periarticular tissues
   C. Bone marrow
   D. Muscle
   E. None of the above

5. RECENT RESEARCH INDICATES THAT THE SEVERAL VARIETIES OF ENCAPSULATED NERVE ENDINGS:
   A. Are restricted to hairy skin
   B. Respond to mechanical deformation
   C. Are served by unmyelinated nerve fibres
   D. Are concerned in temperature transduction
   E. Contain capillary networks

6. THE C-FIBRES OF THE PERIPHERAL PAIN PATH:
   A. Do not end directly upon the cells of origin of the spinothalamic tract
   B. Are antagonised in the posterior grey horn by the inhibitory influence of accompanying A-fibres
   C. Exert "positive feed-forward" by arousal of excitatory internuncials
   D. All of the above
   E. A and B only

7. MEISSNER'S CORPUSCLES ARE:
   A. Categorised as encapsulated nerve endings
   B. The only nerve endings concerned in transduction of touch stimuli
   C. Also known as tendon organs
   D. Served by unmyelinated sensory nerves
   E. A and B only

8. THE RATE OF REGENERATION OF PERIPHERAL NERVES FOLLOWING INJURY IS:
   A. 1–3 micra per day
   B. 1–3 mm. per hour
   C. 1–3 mm. per day
   D. 1–3 mm. per week
   E. 1–3 mm. per month

9. THE SPINAL CORD EXTENDS FROM FORAMEN MAGNUM TO THE LOWER BORDER OF THE ---------- VERTEBRA:
   A. Tenth thoracic
   B. Eleventh thoracic
   C. First lumbar
   D. Third lumbar
   E. Fifth lumbar

10. IN THE POSTNATAL PERIOD THE GREATEST GROWTH IN THE GREY MATTER OF THE CNS IS OF:
    A. Nerve cell numbers
    B. Length of axonal processes
    C. Dendritic trees
    D. Size of perikarya
    E. None of the above

11. THE CELLS OF THE POSTERIOR GREY HORN OF THE SPINAL CORD ARE ARRANGED:
    A. In clusters
    B. In laminae from apex to base
    C. In laminae from medial to lateral
    D. Diffusely
    E. In alternating large and small cell groups

12. GROSSLY THE SPINAL CORD PRESENTS TWO SWELLINGS:
    A. Cervical and thoracic
    B. Cervical and lumbar
    C. Thoracic and lumbar
    D. Thoracic and sacral
    E. Lumbar and sacral

13. THE INTERNAL VERTEBRAL VENOUS PLEXUS OCCUPIES:
    A. The extradural space
    B. The subdural space
    C. The subarachnoid space
    D. The spinal cord
    E. The foramen transversarium

14. IN ELICITING A TENDON REFLEX, THE SEQUENCE OF NEURONES ACTIVATED IS:
    A. Tendon afferent, internuncial, alpha efferent
    B. Spindle afferent, internuncial, alpha efferent
    C. Spindle afferent, internuncial, gamma efferent
    D. Tendon afferent, alpha efferent
    E. Spindle afferent, alpha efferent

15. ELICITATION OF THE KNEE JERK DEPENDS ON:
    A. Muscle spindles in the quadriceps femoris muscle
    B. Golgi tendon organs in the ligamentum patellae
    C. Integrity of the sacral segments of the spinal cord
    D. All of the above
    E. A and B only

16. ELICITATION OF THE ANKLE JERK DEPENDS ON:
    A. Integrity of the sciatic nerve
    B. Reflex contraction of the gastrocnemius muscle
    C. Integrity of internuncial neurones in the spinal cord
    D. All of the above
    E. A and B only

17. THE FOLLOWING IS/ARE CONSIDERED CHARACTERISTIC OF POSTERIOR COLUMN DISEASE (BELOW LEVEL OF LESION):
    A. Impairment of position sense
    B. Impairment of pain sense
    C. Impairment of temperature sense
    D. All of the above
    E. A and B only

18. THE FOLLOWING TRACTS ARISE IN THE DORSAL GREY HORN, *EXCEPT*:
    A. Ventral spinocerebellar
    B. Dorsal spinocerebellar
    C. Anterior spinothalamic
    D. Lateral spinothalamic
    E. Fasciculus gracilis

19. THE CELL IN THE CNS MOST LIKE THE SCHWANN CELL IN FUNCTION IS THE:
    A. Fibrous astrocyte
    B. Protoplasmic astrocyte
    C. Microglial cell
    D. Pericyte
    E. Oligodendrocyte

20. IN THE CNS PORTIONS OF FIRST ORDER SENSORY NEURONES ARE FOUND:
    A. In the posterior columns
    B. In the posterolateral tract of Lissauer
    C. In the lateral columns
    D. In the anterior columns
    E. A and B

21. THE SACRAL SEGMENTS OF THE SPINAL CORD MAY BE CRUSHED BY FRACTURE OF:
    A. The first lumbar vertebra
    B. The third lumbar vertebra
    C. The fifth lumbar vertebra
    D. The first and second sacral vertebrae
    E. None of the above

22. IN THE CENTRAL NERVOUS SYSTEM, THE CELLS RESPONSIBLE FOR THE PRODUCTION OF MYELIN ARE THE:
    A. Astrocytes
    B. Oligodendrocytes
    C. Microglia
    D. Ependymal cells
    E. Cells of Schwann

23. THE SULCUS LIMITANS OF THE DEVELOPING NEURAL TUBE SEPARATES:
    A. Alar lamina from basal lamina
    B. The alar laminae
    C. The basal laminae
    D. Ependymal layer from mantle layer
    E. Mantle layer from marginal layer

24. THE CELL BODIES OF THE NERVE FIBRES MAKING UP FASCICULUS GRACILIS ARE FOUND IN:
    A. Posterior root ganglia
    B. Posterior grey horn
    C. Nucleus gracilis
    D. Lateral grey horn
    E. Anterior grey horn

25. THE SOMATIC EFFERENT CELL COLUMN OF THE PRIMITIVE NEURAL TUBE OCCUPIES THE:
    A. Roof plate
    B. Floor plate
    C. Alar lamina
    D. Basal lamina
    E. Marginal layer

26. SPONGIOBLASTS ARE PRECURSORS OF:
    A. Motor neurones
    B. Sensory neurones
    C. Neuroglial cells
    D. All of the above
    E. A and B only

27. THE CORTICOSPINAL PATHWAY IS INJURED ON THE LEFT SIDE AT C2 LEVEL OF CORD. YOU EXPECT TO FIND:
    A. Weakness of limb muscles on the LEFT side
    B. Spasticity of muscles on the LEFT side
    C. Brisk reflexes on LEFT side
    D. All of the above
    E. None of the above

28. THE CEPHALIC FLEXURE OF THE BRAIN OCCURS IN REGION OF:
    A. Medulla oblongata
    B. Pons
    C. Midbrain
    D. Diencephalon
    E. Telencephalon

29. IN THE UPPER THIRD OF THE MEDULLA OBLONGATA THE MEDIAL LEMNISCUS IS RELATED ON ITS MEDIAL ASPECT TO:
   A. Its opposite number
   B. The medial longitudinal bundle
   C. The central grey matter
   D. The central canal
   E. All of the above

30. THE FOLLOWING FIBRES CROSS THE MIDLINE IN THE MEDULLA OBLONGATA:
   A. Internal arcuate
   B. Corticospinal
   C. Olivocerebellar
   D. All of the above
   E. A and B only

31. A SECTION THROUGH THE ADULT MEDULLA, IN THE MID-OLIVARY REGION, WOULD *NOT* DEMONSTRATE THE:
   A. Hypoglossal nucleus
   B. Pyramidal decussation
   C. Lateral reticular nucleus
   D. Nucleus ambiguus
   E. Dorsal nucleus of vagus

32. THE CISTERNS (CISTERNAE) OF THE BRAIN ARE:
   A. Localized enlargements of the subarachnoid space
   B. Cerebral cysts appearing in the elderly
   C. Synonyms for the lateral ventricles
   D. Synonyms for the cavernous sinuses
   E. None of the above

33. IN THE FOURTH VENTRICLE THE FACIAL COLLICULUS OVERLIES THE ---------- NUCLEUS:
   A. Oculomotor
   B. Trigeminal (motor)
   C. Abducent
   D. Facial
   E. Cochlear

34. BRANCH(ES) OF THE BASILAR ARTERY:
    A. Superior cerebellar
    B. Posterior cerebral
    C. Anterior spinal
    D. Vertebral
    E. A and B

35. THE CEREBELLUM IS AN OUTGROWTH OF THE:
    A. Diencephalon
    B. Mesencephalon
    C. Prosencephalon
    D. Metencephalon
    E. Myelencephalon

36. THE CEREBELLUM SENDS EFFERENT FIBRES TO EACH OF THE FOLLOWING, *EXCEPT:*
    A. The red nucleus of the opposite side
    B. The thalamus of the opposite side
    C. Reticular formation
    D. The vestibular nuclei of the same side
    E. The substantia nigra

37. THE STRUCTURE CLOSEST TO THE CRUS CEREBRI IS THE:
    A. Substantia nigra
    B. Red nucleus
    C. Medial lemniscus
    D. Opposite crus
    E. Superior cerebellar peduncle

38. THE RED NUCLEUS IS LOCATED IN THE PATH OF THE:
    A. Medial lemniscus
    B. Medial longitudinal fasciculus
    C. Fasciculus gracilis
    D. Superior cerebellar peduncle
    E. Inferior cerebellar peduncle

39. THE CEREBRAL AQUEDUCT (OF SYLVIUS) CONNECTS:
    A. The lateral ventricles
    B. Lateral and third ventricles
    C. Third and fourth ventricles
    D. Fourth ventricle and subarachnoid space
    E. Fourth ventricle and central canal of spinal cord

40. THE FOLLOWING STRUCTURES LIE IN THE FLOOR OF THE THIRD VENTRICLE:
    A. Optic chiasma, infundibulum
    B. Anterior commissure, infundibulum
    C. Anterior commissure, optic chiasma
    D. Hypothalamus
    E. Hypothalamus, infundibulum

41. A COMPONENT OF THE SUPERIOR CEREBELLAR PEDUNCLE:
    A. Dorsal spinocerebellar tract
    B. Ventral spinocerebellar tract
    C. Corticoponticerebellar tract
    D. Olivocerebellar tract
    E. External arcuate fibres

42. THE FORAMEN OF MONRO CONNECTS:
    A. The two lateral ventricles
    B. Lateral ventricle with third ventricle
    C. Third and fourth ventricles
    D. Fourth ventricle with subarachnoid space
    E. Tela choroidea with third ventricle

43. THE FOLLOWING NUCLEI BELONG TO THE BASAL GANGLIA, *EXCEPT:*
    A. The claustrum
    B. The substantia nigra
    C. The amygdaloid nucleus
    D. The caudate nucleus
    E. The lentiform nucleus

44. BETZ CELLS CONSTITUTE THE FOLLOWING PERCENTAGE OF THE CORTICOSPINAL TRACT NEURONES:
    A. 0
    B. 2.5
    C. 10
    D. 50
    E. 100

45. THE MAIN SITE OF TERMINATION OF THE RIGHT SPINO-THALAMIC TRACT IS:
    A. Bilaterally, in the brain stem
    B. In the right thalamus
    C. In the left thalamus
    D. In the spinal cord
    E. In the tectum of the midbrain

46. THE CRUS CEREBRI CONTAINS:
    A. Medial lemniscus
    B. Spinothalamic tract
    C. Temporopontine fibres
    D. Lateral lemniscus
    E. None of the above

47. DEVELOPMENTALLY, THE CORPUS STRIATUM BELONGS TO THE:
    A. Telencephalon
    B. Diencephalon
    C. Mesencephalon
    D. Metencephalon
    E. Myelencephalon

48. FOLDS OF THE CEREBRAL CORTEX OVERLYING THE INSULA ARE CALLED THE:
    A. Temporal lobes
    B. Corpus callosum
    C. Opercula
    D. Commissures of the fornix
    E. Pallia

49. NYSTAGMUS DEPENDS UPON CONNEXIONS BETWEEN VESTIBULAR AND OCULOMOTOR NUCLEI VIA:
    A. Medial longitudinal bundle
    B. Medial lemniscus
    C. Reticular formation
    D. Spinal lemniscus
    E. Central grey matter

50. IN EXECUTING THE FINGER-NOSE-TEST, AN "INTENTION TREMOR" — WHICH IS PRESENT WHETHER THE EYES ARE OPEN OR CLOSED — SUGGESTS DISORDER OF:
    A. Upper motor neurones
    B. Lower motor neurones
    C. The extrapyramidal system
    D. The cerebellar pathway
    E. None of the above

51. THE LAST DESCENDING TRACT TO UNDERGO MYELINATION:
    A. Rubrospinal
    B. Tectospinal
    C. Olivospinal
    D. Corticospinal
    E. Reticulospinal

52. THE RETROLENTIFORM (RETROLENTICULAR) PART OF THE INTERNAL CAPSULE CONTAINS:
    A. The optic radiation
    B. The pyramidal tract
    C. The main thalamocortical system
    D. Frontopontine fibres
    E. Temporopontine fibres

53. THE TRIANGULAR AREA BETWEEN FORNIX AND CORPUS CALLOSUM IS OCCUPIED BY THE:
    A. Insula
    B. Septum pellucidum
    C. Genu
    D. Uncus
    E. Lateral geniculate body

54. THE INTERNAL CAPSULE OF THE BRAIN LIES BETWEEN:
    A. The caudate nucleus and thalamus medially, and the lentiform nucleus laterally
    B. The thalamus medially and the caudate and lentiform nuclei laterally
    C. The thalamus and lentiform nucleus medially, and the caudate nucleus laterally
    D. The thalamus medially and the basal ganglia laterally
    E. None of the above

55. THE FOLLOWING FIBRE SYSTEMS RELAY IN THE THALAMUS:
    A. Medial lemniscus
    B. Superior cerebellar peduncle
    C. Temporopontine
    D. All of the above
    E. A and B only

56. CEREBRAL COMMISSURE(S):
    A. Corpus callosum
    B. Crus cerebri
    C. Internal capsule
    D. External capsule
    E. A and B

57. BROCA'S AREA OCCUPIES THE:
    A. Superior frontal gyrus
    B. Middle frontal gyrus
    C. Inferior frontal gyrus
    D. Superior temporal gyrus
    E. Inferior temporal gyrus

58. THE PARIETAL LOBE IS OCCUPIED BY THE FOLLOWING CONVOLUTIONS, *EXCEPT*:
    A. Postcentral
    B. Supramarginal
    C. Angular
    D. Superior parietal
    E. Lingual

59. THE CELLS OF ORIGIN OF THE OPTIC RADIATION OCCUPY THE:
    A. Ganglion cell layer of retina
    B. Medial geniculate body
    C. Lateral geniculate body
    D. Superior colliculus
    E. Occipital cortex

60. THE CORTICOBULBAR SYSTEM:
    A. Traverses the internal capsule
    B. Contains the upper motor neurones of supply to cranial nerves
    C. Terminates bilaterally in relation to the oculomotor nuclei
    D. All of the above
    E. A and B only

61. THE SPASTICITY FOLLOWING HAEMORRHAGE INTO THE INTERNAL CAPSULE OF THE BRAIN IS THOUGHT TO BE DUE TO INVOLVEMENT OF THE ---------- FIBRE SYSTEM:
    A. Corticospinal
    B. Corticoreticular
    C. Corticobulbar
    D. Corticoponticerebellar
    E. Corticothalamic

62. THE MAIN MOTOR AREA (BRODMAN'S AREA 4) ON THE CEREBRAL CORTEX IS LOCATED IN THE:
    A. Frontal lobe
    B. Parietal lobe
    C. Occipital lobe
    D. Temporal lobe
    E. Lips of the lateral sulcus

63. BURIED IN THE LATERAL SULCUS OF THE CEREBRAL HEMISPHERE IS THE:
    A. Occipital lobe
    B. Splenium
    C. Operculum
    D. Insula
    E. Pars triangularis

64. NEUROLOGICAL EXAMINATION OF A PATIENT REVEALS RIGHT-SIDED HEMIPARESIS, INCREASED TENDON REFLEXES, HOMONYMOUS HEMIANOPIA AND POSITIVE BABINSKI SIGN. WHICH OF THE FOLLOWING IS THE MOST LIKELY SITE OF THE LESION?
    A. Left occipital cortex
    B. Left frontal lobe
    C. Left internal capsule
    D. Optic chiasma
    E. Right thalamus

65. THE ABOVE PATIENT'S SYMPTOMATOLOGY WOULD LEAD YOU TO EXPECT:
    A. Some right-sided sensory deficit
    B. Increased abdominal reflexes
    C. Cerebellar signs
    D. Abnormal pupillary reactions
    E. A and B

66. A CRANIAL NERVE LESION CORRESPONDING TO AN UPPER MOTOR NEURONE LESION OF A SPINAL NERVE IS CALLED:
    A. Suprasegmental
    B. Supratentorial
    C. Supranuclear
    D. Supramarginal
    E. Supraorbital

67. BIPOLAR NERVE CELLS ARE FOUND IN:
   A. Retina
   B. Olfactory epithelium
   C. Cochlea
   D. All of the above
   E. A and B only

68. THE GANGLIONIC CELLS OF THE RETINA ARE HOMOLOGOUS WITH:
   A. Posterior root ganglion neurones
   B. Posterior grey horn neurones
   C. Anterior grey horn neurones
   D. Autonomic ganglia
   E. None of the above

69. THE FOLLOWING STRUCTURES ARE FOUND AT THE LEVEL OF THE NUCLEUS OF THE OCULOMOTOR NERVE, *EXCEPT*:
   A. Red nucleus
   B. Pretectal nucleus
   C. Substantia nigra
   D. Decussation of the superior cerebellar peduncles
   E. Superior colliculus

70. SUPERFICIAL ORIGIN OF TROCHLEAR NERVE:
   A. Anterior surface of midbrain
   B. Lateral surface of midbrain
   C. Posterior surface of midbrain
   D. Lower border of pons
   E. Posterior surface of pons

71. IN THE CAVERNOUS SINUS THE NERVE CLOSEST TO THE INTERNAL CAROTID ARTERY IS THE:
   A. Oculomotor
   B. Trochlear
   C. Ophthalmic
   D. Maxillary
   E. Sympathetic plexus

72. THE ABDUCENT NERVE IS ESPECIALLY LIABLE TO BE COMPRESSED BY A RISE IN INTRACRANIAL PRESSURE BECAUSE:
    A. It emerges at the lower border of the pons
    B. It is related to the basisphenoid
    C. It crosses the sharp apex of the petrous temporal bone
    D. It traverses the superior orbital fissure
    E. It does not traverse the subarachnoid space

73. IN WHICH MEDIO-LATERAL ORDER DO FACIAL, ABDUCENT AND VESTIBULOCOCHLEAR NERVES ARISE FROM THE LOWER BORDER OF THE PONS?
    A. VIII, VII, VI
    B. VIII, VI, VII
    C. VII, VIII, VI
    D. VI, VII, VIII
    E. VI, VIII, VII

74. FIBRES FROM THE DORSAL COCHLEAR NUCLEUS JOIN FIBRES FROM THE VENTRAL COCHLEAR NUCLEUS IN THE:
    A. Nucleus of the vestibular nerve
    B. Trapezoid body
    C. Medial longitudinal fasciculus
    D. Medial lemniscus
    E. None of the above

75. AXONS FROM THE NUCLEUS AMBIGUUS INNERVATE MUSCLES OF THE:
    A. Eye
    B. Tongue
    C. Larynx
    D. Ear
    E. Jaw

76. THE NUCLEUS AMBIGUUS GIVES ORIGIN TO MOTOR FIBRES THAT RUN THROUGH THE ————————— NERVES:
    A. Vagus, trigeminal and facial
    B. Glossopharyngeal and vagus
    C. Vagus, hypoglossal and facial
    D. Facial, abducent and oculomotor
    E. Trigeminal, abducent and facial

QUESTIONS 77–101:

THE SET OF LETTERED HEADINGS BELOW IS FOLLOWED BY A LIST OF NUMBERED WORDS OR PHRASES. FOR EACH NUMBERED WORD OR PHRASE SELECT THE CORRECT ANSWER UNDER:

    A.   If the item is associated with A only
    B.   If the item is associated with B only
    C.   If the item is associated with both A and B
    D.   If the item is associated with neither A nor B

    A.   Anterior cerebral artery
    B.   Middle cerebral artery
    C.   Both
    D.   Neither

77. Blood supply to "leg area" of motor cortex
78. Blood supply to full anteroposterior length of the internal capsule
79. Blood supply to auditory cortex
80. Blood supply to visual cortex
81. In contact with corpus callosum

    A.   Red nucleus
    B.   Lentiform nucleus
    C.   Both
    D.   Neither

82. Part of "basal ganglia"
83. Part of "corpus striatum"
84. Afferents received from cerebral cortex
85. Efferents pass to thalamus
86. Blood supply from middle cerebral artery

A. Unipolar nerve cells
B. Multipolar nerve cells
C. Both
D. Neither

87. Found in sympathetic ganglia
88. Found in dorsal grey horn of spinal cord
89. Restricted to sensory pathways
90. Usually no synapse in vicinity
91. Derived from embryonic bipolar neuroblasts

A. Right fasciculus gracilis
B. Right fasciculus cuneatus
C. Both
D. Neither

92. Proprioception from right upper limb
93. Temperature sensation from right lower limb
94. Parent cells in posterior grey horn
95. Termination in medulla oblongata
96. Crosses midline before ascending

A. Upper motor neurone lesion
B. Lower motor neurone lesion
C. Both
D. Neither

97. Characterised by rapid wasting
98. Characterised by fasciculation
99. Characterised by weakness
100. Characterised by loss of reflexes
101. Characterised by Babinski sign

QUESTIONS 102–125:

DIRECTIONS: In the following series of questions, one or more of the four items is/are correct. Answer A if 1, 2, 3 are correct; B if 1 and 3 are correct; C if 2 and 4 are correct; D if only 4 is correct; and E if all four are correct.

102. EPENDYMA:
    1. Is composed of stratified squamous epithelium
    2. Lines walls of lateral ventricles
    3. Is of mesodermal origin
    4. Contains ciliated cells

103. ARACHNOID GRANULATIONS:
    1. Are most numerous in the region of the superior sagittal sinus
    2. Diminish in size with advancing age
    3. Create perforations in the dura mater of venous sinuses
    4. Are the sole portals of entry of cerebrospinal fluid into the general circulation

104. THE TECTUM OF THE MIDBRAIN:
    1. Lies dorsal to the aqueduct
    2. Receives the medial lemniscus
    3. Contains the four colliculi
    4. Is occupied by reticular formation

105. THE NODES OF RANVIER LOCATED ALONG NERVE FIBRES ARE:
    1. Sites of nerve branching
    2. Myelinated
    3. Important in saltatory conduction
    4. Synonymous with the clefts of Schmidt-Lantermann

106. LOWER MOTOR NEURONE LESIONS ARE CHARACTERISED BY:
    1. Flaccidity
    2. Loss of reflexes
    3. Wasting
    4. Sensory loss

107. THE LATERAL SPINOTHALAMIC TRACT SUBSERVES:
    1. Heat
    2. Pain
    3. Cold
    4. Touch

108. ANNULOSPIRAL NERVE ENDINGS ARE ACTIVATED BY:
    1. Contraction of intrafusal muscle fibres
    2. Stretch of antagonist muscles
    3. Stretch of related extrafusal muscle fibres
    4. Contraction of related extrafusal muscle fibres

109. A LEFT HOMONYMOUS HEMIANOPIA CAN BE PRODUCED BY A LESION IN THE RIGHT:
    1. Occipital cortex
    2. Optic radiation
    3. Optic tract
    4. Optic nerve

110. BILATERAL REPRESENTATION ON THE CEREBRAL CORTEX:
    1. Hearing
    2. Motor innervation of lower face
    3. Ocular movements
    4. Movements of tongue

111. WHEN THE SPINAL CORD IS FULLY DEVELOPED, THE EMBRYOLOGICAL MANTLE LAYER PERSISTS AS THE:
    1. Neural crest
    2. White matter
    3. Ependymal layer
    4. Grey matter

112. THE NEURAL CREST OF THE EMBRYO GIVES RISE TO:
1. The posterior root ganglia of the spinal nerves
2. Schwann cells
3. Melanoblasts
4. Chromaffin cells

113. GLIAL CELLS OF ECTODERMAL ORIGIN ARE:
1. Ependymal cells
2. Astrocytes
3. Oligodendroglia
4. Microglia

114. SYNAPTIC BOUTONS MAY BE FOUND ON THE:
1. Soma
2. Axon hillock
3. Dendrites
4. Initial unmyelinated segment of the axon

115. EXPERIMENTAL DESTRUCTION OF THE PYRAMID IN MONKEYS IS KNOWN TO PRODUCE:
1. Paralysis of the last four cranial nerves
2. Paralysis of the limbs on the opposite side
3. A general fall in vasomotor tone
4. Flaccidity of the limbs on the opposite side

116. ELICITATION OF A TENDON REFLEX DEPENDS ON THE INTEGRITY OF RELATED:
1. Muscle spindles
2. Dorsal nerve roots
3. Ventral nerve roots
4. Tendon organs

117. LEFT-SIDED HEMISECTION OF SPINAL CORD IN THE MID-THORACIC REGION MAY PRODUCE:
1. Loss of pain sensation in the left leg
2. Loss of pain sensation in the right leg
3. Loss of vibration sense in the right leg
4. Loss of vibration sense in the left leg

118. THE NUCLEUS AMBIGUUS:
    1. Innervates striated muscle of branchial origin
    2. Contributes to the facial nerve
    3. Occupies the lateral part of the medulla oblongata
    4. Is usually absent in man

119. THE PRETECTAL NUCLEUS IS ON THE PATHWAY OF THE ────────── REFLEX:
    1. Light
    2. Accommodation
    3. Consensual
    4. Convergence

120. THE SPINAL CORD IS CONTINUOUS INFERIORLY WITH THE FILUM TERMINALE WHICH:
    1. Continues caudally from the tip of the conus medullaris
    2. Begins at S-2
    3. Is attached to the coccyx
    4. Consists of arachnoid and dura

121. UPPER MOTOR NEURONE LESIONS ARE CHARACTERISED BY:
    1. Spasticity
    2. Increased tendon reflexes
    3. Babinski sign
    4. Absent abdominal reflexes

122. A PATIENT SUFFERS FROM PTOSIS OF HER RIGHT EYELID, TOGETHER WITH DIPLOPIA (DOUBLE VISION). FURTHER EXAMINATION OF THIS EYE IS LIKELY TO REVEAL:
    1. Weakness of abduction
    2. Convergent squint
    3. Loss of corneal sensation
    4. Dilatation of the pupil

123. THE CENTRAL GREY MATTER OF THE MIDBRAIN:
    1. Surrounds the aqueduct
    2. Contains the Edinger-Westphal nucleus
    3. Contains the nucleus of the trochlear nerve
    4. Contains the red nucleus

124. DESTRUCTION OF THE RIGHT OPTIC TRACT WILL BE FOLLOWED BY:
   1. Massive Wallerian degeneration in the right optic nerve
   2. Loss of the right corneal reflex
   3. Blindness in the right visual field
   4. Blindness in the left visual field

125. THE RETICULAR FORMATION:
   1. Occupies medulla, pons and midbrain
   2. Is played upon by fibres from the cerebral cortex
   3. Is influenced by the cerebellum
   4. Is concerned in reflexes regulating muscle tone

QUESTIONS 126–140:

THE GROUP OF QUESTIONS BELOW CONSISTS OF FIVE LETTERED HEADINGS, FOLLOWED BY A LIST OF NUMBERED PHRASES. FOR EACH NUMBERED PHRASE SELECT THE *ONE* HEADING WHICH IS MOST CLOSELY RELATED TO IT.
NOTE: EACH CHOICE MAY BE USED *ONLY ONCE.*

126. Between pons and middle cerebellar peduncle
127. Between pons and pyramid
128. Between pyramid and olive
129. Between olive and inferior cerebellar peduncle
130. Between pons and olive

   A. Trigeminal nerve
   B. Abducent nerve
   C. Facial nerve
   D. Glossopharyngeal nerve
   E. Hypoglossal nerve

131. Unipolar nerve cells
132. Preganglionic sympathetic nerve fibres
133. Multipolar neurones
134. Cervical plexus
135. Postganglionic sympathetic nerve fibres

   A. Posterior nerve roots
   B. Anterior horn cells
   C. Grey rami communicantes
   D. Ventral rami
   E. Anterior nerve roots

136. Arm area of motor cortex
137. Visuosensory cortex
138. Leg area of sensory cortex
139. Medulla oblongata
140. Pons

A. Anterior cerebral artery
B. Middle cerebral artery
C. Posterior cerebral artery
D. Basilar artery
E. Posterior inferior cerebellar artery

QUESTIONS 141–230:

IN REPLY TO THE FOLLOWING QUESTIONS INDICATE WHETHER YOU THINK EACH STATEMENT IS *TRUE* OR *FALSE:*

UNMYELINATED NERVE FIBRES ARE:
141. Absent from ventral nerve roots
142. Devoid of Schwann cell investment
143. Characteristic of preganglionic neurones
144. Devoid of nodes of Ranvier
145. Confined to the grey matter where they occur within the CNS

GOLGI TENDON ORGANS:
146. Have a higher threshold of excitation than muscle spindles
147. Contain nerve endings in direct contact with collagen filaments
148. May be activated by rapid passive lengthening or by strong contraction of the related muscle
149. Their afferent fibres have monosynaptic connections with alpha efferents to the parent muscle
150. Their afferents have monosynaptic connections with alpha efferents to antagonist muscles

FOLLOWING SECTION OF A PERIPHERAL NERVE:
151. The endoneurium in the distal stump undergoes a gradual increase in thickness
152. Myelin sheaths in the distal stump break up and are phagocytosed
153. Axis cylinders break up simultaneously throughout the entire length of the distal stump
154. Schwann cells migrate from both cut surfaces in an attempt to bridge the gap
155. Degenerative changes in the proximal stump only occur distal to the final node of Ranvier

REGENERATING PERIPHERAL NERVES:
156. Require the support of living Schwann cells in the peripheral stump
157. Grow at the same rate in motor and sensory nerves of similar size
158. In muscle, re-establish neuromuscular junctions at the original locations
159. Enter the distal stump in large numbers provided apposition of cut ends is good
160. Coarse fibres acquire new myelin sheaths as they traverse the distal stump

GAMMA MOTONEURONES:
161. Supply only intrafusal muscle fibres
162. Stimulation causes contraction of intrafusal muscle fibres
163. Contribute the small-fibre myelinated axon component to the anterior nerve root
164. Are predominantly influenced by impulses from the reticular formation
165. Their axons enter the muscle in the so-called "motor nerve"

THE FOLLOWING ULTRASTRUCTURAL FEATURES CHARACTERIZE A SYNAPSE:
166. Thickening of the apposed plasma membranes
167. The presence of clear vesicles in the axon terminal
168. The presence of dense-core vesicles in the adjacent region of dendrite or soma
169. Intrusion of delicate glial processes into the synaptic cleft
170. Tight junctions at the margins of the synaptic cleft

THE SUBSTANTIA GELATINOSA:
171. Is composed of neurones deficient in neurofilaments
172. Belongs to the pathway for pain
173. Does not belong to the posterior grey horn
174. Receives the terminals of Lissauer's tract
175. Axosomatic synapses therein outnumber axodendritic

COMPLETE TRANSECTION OF THE THORACIC CORD IS CHARACTERIZED BY:
176. Permanent loss of voluntary control of lower limb muscles
177. Permanent loss of tendon reflexes in the lower limb
178. Permanent loss of sensation below the level of the lesion
179. Return of voluntary sphincteric control
180. Impotence

REGENERATION IN THE SPINAL CORD IS INEFFECTIVE BECAUSE OF:
181. Lack of endoneurial sheaths
182. Occupation of vacated synaptic sites by neighbouring nerve sprouts
183. Obstruction by neuroglial scars
184. Relatively poor blood supply of C.N.S.
185. Large scale death of neurones whose axons are injured

THE FASCICULUS GRACILIS:
186. Consists of first order neurones
187. Contains neurones extending without interruption from sole of foot to brain stem
188. Lies lateral to fasciculus cuneatus
189. Would be unaffected by destruction of lumbosacral posterior root ganglia
190. Is entirely composed of uncrossed fibres

RECENT RESEARCH INDICATES THAT THE POSTERIOR COLUMNS:
191. Are a slowly conducting afferent system
192. Are important for two-point discrimination
193. Are the only pathway for joint sense
194. Have a high degree of somatotopic organization
195. Provide advance information about body position required for smooth execution of movements

THE SUBSTANTIA NIGRA:
196. Is restricted to the midbrain
197. Lies between cerebral peduncle and tegmentum
198. Projects to ventrolateral nucleus of thalamus
199. Appears to be mainly inhibitory in function
200. Is composed of monoaminergic neurones

THE HYPOTHALAMUS:
201. Occupies the side wall of the third ventricle
202. Receives afferents directly from the cerebral cortex
203. Communicates with the opposite hypothalamus by commissural fibres
204. Contributes fibres which emerge in cranial nerves which contain autonomic axons
205. Receives afferents from the amygdaloid nucleus

THE MOTOR CORTEX:
206. Lies entirely behind the central sulcus
207. Extends onto the medial surface of the hemisphere
208. Receives afferents from the ventrolateral nucleus of thalamus
209. Is linked by association fibres with the sensory cortex
210. Gives origin to 90% of the pyramidal tract

IN THE INTERNAL CAPSULE:
211. The lentiform nucleus is related to the anterior and posterior limbs
212. The optic radiation occupies the retrolentiform portion
213. The anterior limb contains only descending fibres
214. Only a minority of the fibres are corticospinal
215. Blood supply is exclusively from the middle cerebral artery

THE CENTRAL ARTERY OF THE RETINA:
216. Enters the eyeball within the optic nerve
217. Enters the retina near the centre of the optic disc
218. Usually gives rise to four main branches
219. Anastomoses with no other artery
220. Its branches and trunk are accompanied by veins

THE OPTIC NERVE:
221. Comprises the central processes of the bipolar cell layer of the retina
222. Is ensheathed throughout its length by the meninges
223. Terminates in the occipital cortex
224. Is homologous with the posterior nerve root of a spinal nerve
225. Passes through the superior orbital fissure

THE PYRAMIDAL TRACT:
226. Is synonymous with the corticospinal tract
227. Occupies the tegmentum of the midbrain
228. Is so named because it makes up the pyramid in the medulla oblongata
229. Contains about one million nerve fibres
230. Terminates mainly on internuncial neurones

SECTION OF PERIPHERAL NERVE

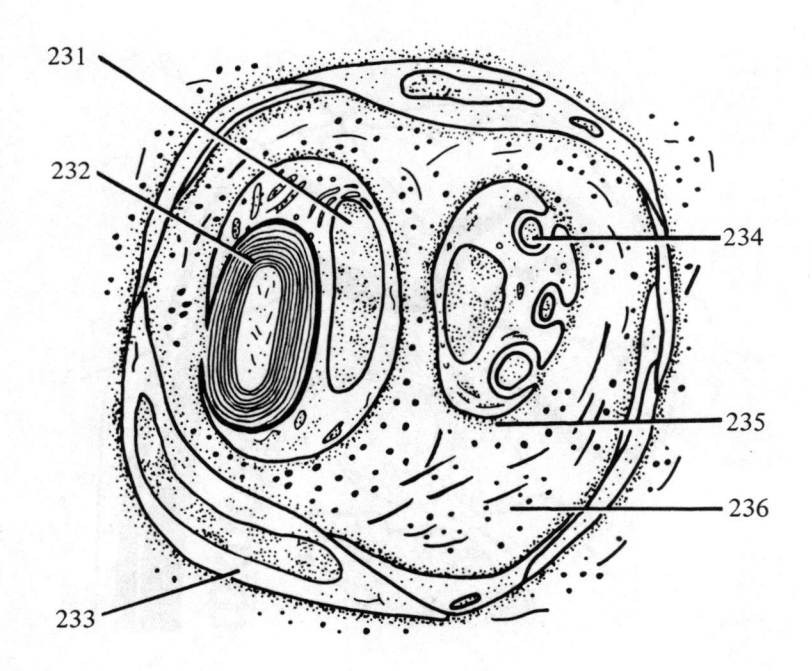

IDENTIFY THE NUMBERED STRUCTURES:

- A. Perineurium
- B. Mesaxon
- C. Schwann cell nucleus
- D. Elastic coat
- E. Endoneurium
- F. Myelin sheath
- G. Basal lamina
- H. Axon
- I. Capillary
- J. Epineurium
- K. Neurokeratin
- L. Fibroblast

163

MEDIAN SECTION OF HEAD

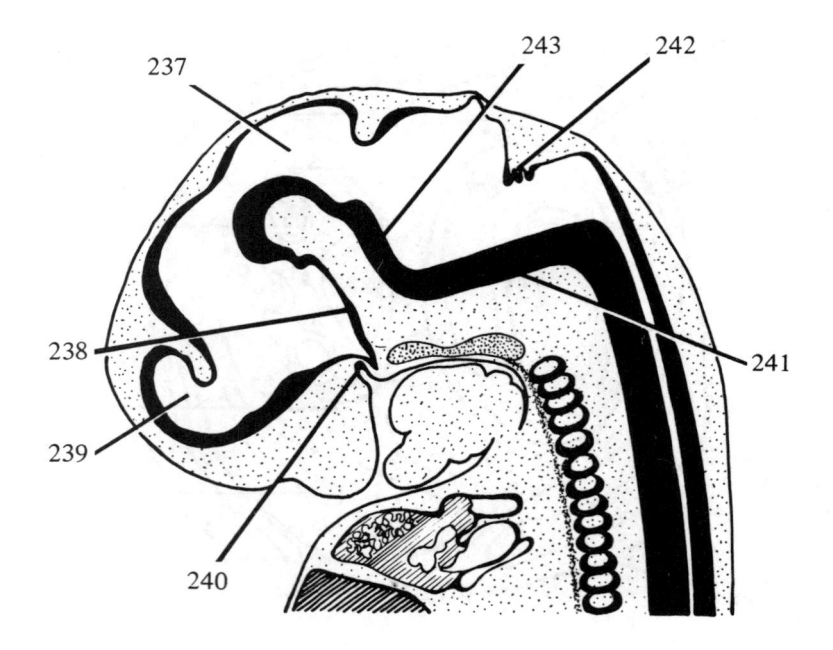

IDENTIFY THE NUMBERED STRUCTURES:

A.	Adenohypophysis	G.	Lateral ventricle
B.	Brain plate	H.	Fourth ventricle
C.	Aqueduct	I.	Third ventricle
D.	Diencephalon	J.	Metencephalon
E.	Basal ganglia	K.	Choroid plexus
F.	Myelencephalon	L.	Cerebellum

164

SECTION OF BRAIN STEM

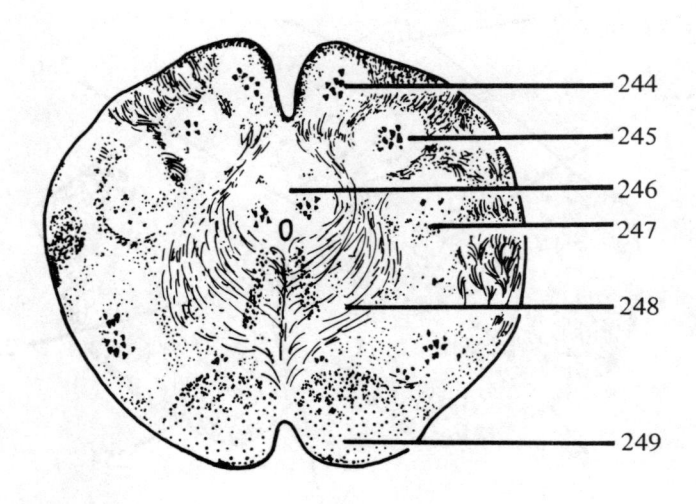

IDENTIFY THE NUMBERED STRUCTURES:

- 244
- 245
- 246
- 247
- 248
- 249

A. Internal arcuate fibres
B. Nucleus cuneatus
C. Olive
D. Central grey matter
E. Lateral reticular nucleus
F. Hypoglossal nucleus
G. Nucleus gracilis
H. Pyramid
I. Spinocerebellar fibres
J. Fasciculus cuneatus
K. Spinal nucleus of fifth nerve
L. Nucleus ambiguus

SECOND SECTION OF BRAIN STEM

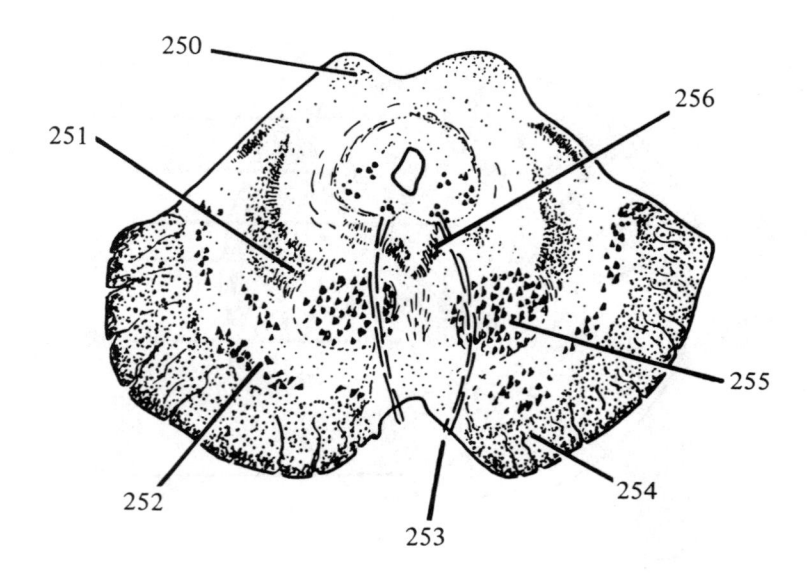

IDENTIFY THE NUMBERED STRUCTURES:

A. Substantia nigra
B. Frontopontine tract
C. Oculomotor nerve
D. Inferior colliculus
E. Gudden's commissure
F. Central tegmental tract
G. Trochlear nerve
H. Superior colliculus
I. Central grey matter
J. Abducent nerve
K. Medial longitudinal bundle
L. Red nucleus
M. Diencephalon
N. Medial lemniscus

MEDIAL SURFACE OF BRAIN

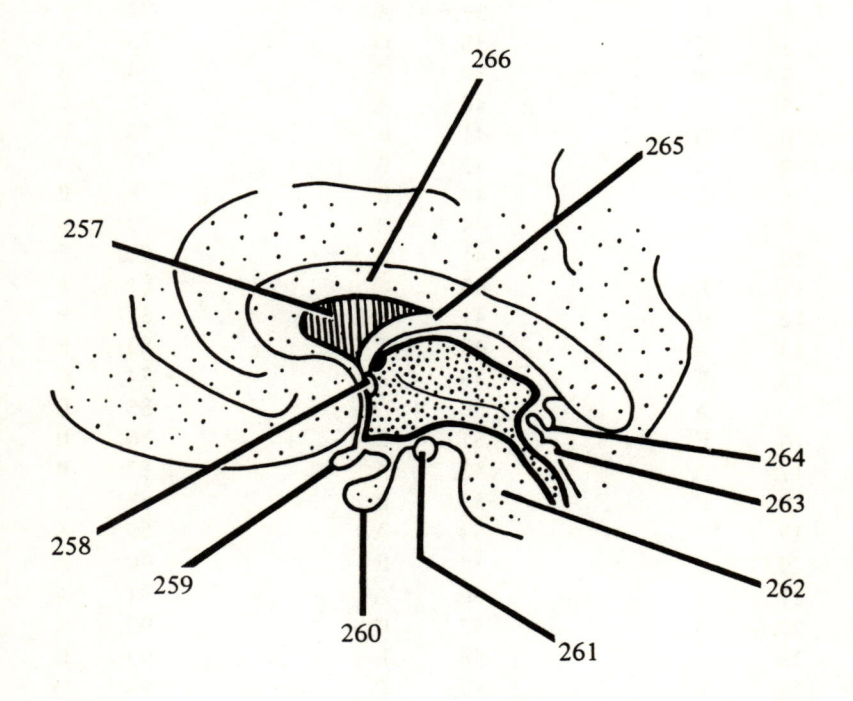

IDENTIFY THE NUMBERED STRUCTURES:

A.	Inferior colliculus	I.	Posterior commissure
B.	Septum pellucidum	J.	Anterior commissure
C.	Pineal gland	K.	Tuber cinereum
D.	Pons	L.	Corpus callosum
E.	Fornix	M.	Mammillary body
F.	Medulla oblongata	N.	Optic chiasma
G.	Cerebellum	O.	Induseum griseum
H.	Olfactory bulb	P.	Hypophysis

ANSWERS

1.	B	36.	E	71.	E
2.	C	37.	A	72.	C
3.	E	38.	D	73.	D
4.	B	39.	C	74.	B
5.	B	40.	A	75.	C
6.	D	41.	B	76.	B
7.	A	42.	B	77.	A
8.	C	43.	B	78.	B
9.	C	44.	B	79.	B
10.	C	45.	A	80.	D
11.	B	46.	C	81.	A
12.	B	47.	A	82.	B
13.	A	48.	C	83.	B
14.	E	49.	A	84.	C
15.	A	50.	D	85.	C
16.	E	51.	D	86.	B
17.	A	52.	A	87.	B
18.	E	53.	B	88.	B
19.	E	54.	A	89.	A
20.	E	55.	E	90.	A
21.	A	56.	A	91.	A
22.	B	57.	C	92.	B
23.	A	58.	E	93.	D
24.	A	59.	C	94.	D
25.	D	60.	D	95.	C
26.	C	61.	B	96.	D
27.	D	62.	A	97.	B
28.	C	63.	D	98.	B
29.	A	64.	C	99.	C
30.	D	65.	A	100.	B
31.	B	66.	C	101.	A
32.	A	67.	D	102.	C
33.	C	68.	B	103.	B
34.	E	69.	D	104.	B
35.	D	70.	C	105.	B

106.	A	145.	F	184.	F
107.	A	146.	T	185.	T
108.	B	147.	T	186.	T
109.	A	148.	T	187.	T
110.	B	149.	F	188.	F
111.	D	150.	F	189.	F
112.	E	151.	T	190.	T
113.	A	152.	T	191.	F
114.	E	153.	T	192.	T
115.	C	154.	T	193.	F
116.	A	155.	T	194.	T
117.	C	156.	T	195.	T
118.	B	157.	T	196.	T
119.	B	158.	T	197.	T
120.	B	159.	T	198.	T
121.	E	160.	T	199.	T
122.	D	161.	T	200.	F
123.	A	162.	T	201.	T
124.	D	163.	T	202.	T
125.	E	164.	T	203.	T
126.	A	165.	T	204.	F
127.	B	166.	T	205.	T
128.	E	167.	T	206.	F
129.	D	168.	F	207.	T
130.	C	169.	F	208.	T
131.	A	170.	F	209.	T
132.	E	171.	T	210.	F
133.	B	172.	T	211.	T
134.	D	173.	F	212.	T
135.	C	174.	T	213.	F
136.	B	175.	F	214.	T
137.	C	176.	T	215.	F
138.	A	177.	F	216.	T
139.	E	178.	T	217.	T
140.	D	179.	F	218.	T
141.	T	180.	F	219.	T
142.	F	181.	T	220.	T
143.	F	182.	T	221.	F
144.	T	183.	T	222.	T

223.	F	238.	D	253.	C
224.	F	239.	G	254.	B
225.	F	240.	A	255.	L
226.	T	241.	F	256.	K
227.	F	242.	K	257.	B
228.	T	243.	J	258.	J
229.	T	244.	G	259.	N
230.	T	245.	B	260.	P
231.	C	246.	D	261.	M
232.	F	247.	K	262.	D
233.	A	248.	A	263.	A
234.	H	249.	H	264.	C
235.	G	250.	H	265.	E
236.	E	251.	N	266.	L
237.	C	252.	A		

SECTION VIII — GENERAL HISTOLOGY

QUESTIONS 1—4:

FOR EACH OF THE FOLLOWING MULTIPLE CHOICE QUESTIONS SELECT THE *ONE* MOST APPROPRIATE ANSWER:

1. CILIATED PSEUDOSTRATIFIED COLUMNAR EPITHELIUM:
    A. Characterises the respiratory portion of the respiratory system
    B. Lines the lower segment of the descending colon
    C. Is rarely associated with goblet cells
    D. Filters and moistens inspired air
    E. Is devoid of a basement membrane in order to facilitate ion transfer to underlying blood vessels

2. STRATIFIED SQUAMOUS EPITHELIUM:
    A. Is protective rather than secretory or absorptive
    B. Consists of a single layer of flattened cells with bulging nuclei
    C. Is readily penetrated by water and ions
    D. Is keratinized where it lines enclosed areas such as the urinary bladder
    E. Has a rich supply of capillaries running between its cell layers

3. ARTERIOLES AND CAPILLARIES DIFFER AS FOLLOWS:
    A. Arterioles have no adventitia
    B. Capillaries have no muscle coat
    C. Capillaries have no intimal layer
    D. Only difference is diameter of lumen
    E. None of the above

4. BLOOD SINUSOIDS:
   A. Differ from capillaries in having wider lumens and incomplete linings
   B. Are found in the kidney and lung
   C. Are absent from red bone marrow
   D. Have smooth muscle cells in their walls
   E. Are rarely lined by reticuloendothelial cells

QUESTIONS 5–34:

THE SET OF LETTERED HEADINGS BELOW IS FOLLOWED BY A LIST OF NUMBERED WORDS OR PHRASES. FOR EACH NUMBERED WORD OR PHRASE SELECT THE CORRECT ANSWER UNDER:

   A. If the item is associated with A only
   B. If the item is associated with B only
   C. If the item is associated with both A and B
   D. If the item is associated with neither A nor B

   A. Aorta
   B. Arterioles
   C. Both
   D. Neither

5. Rich in elastic fibres
6. Muscular media
7. Lined by flat endothelium
8. Rich efferent sympathetic innervation
9. Adventitial coat is present

A. Lymphatic vessels
B. Veins
C. Both
D. Neither

10. Commence blindly
11. Contain valves
12. Lined by endothelium
13. Contain erythrocytes
14. Strengthened by cartilaginous plates

A. Skeletal muscle fibres
B. Cardiac muscle fibres
C. Both
D. Neither

15. Central nuclei
16. Mitochondria are arranged in rows between myofibrils
17. Stimulated via motor end plates
18. Commonly branched
19. Sarcoplasmic continuity by narrow side connections

A. Cartilage
B. Bone
C. Both
D. Neither

20. Rich blood supply
21. Excellent reparative power
22. Collagen in matrix
23. Ground substance rich in mucopolysaccharides
24. Cells commonly occur in pairs or tetrads

A. Collagen fibrils
B. Elastic fibres
C. Both
D. Neither

25. May double their length on being stretched
26. Branch freely
27. Have high resistance to tensile strain
28. Show marked periodicity at 64 nm
29. Retain continuity with parent connective tissue cells

A. Epidermis
B. Dermis
C. Both
D. Neither

30. Origin from mesoderm
31. Gives rise to sweat glands
32. Contains blood vessels
33. Produces keratin
34. Contains lymph vessels

QUESTIONS 35–37:

DIRECTIONS: In the following series of questions, one or more of the four items is/are correct. Answer A if 1, 2, 3 are correct; B if 1 and 3 are correct; C if 2 and 4 are correct; D if only 4 is correct; and E if all four are correct.

35. THE MYOFIBRILS OF VOLUNTARY MUSCLE:
    1. Contain actin
    2. Branch
    3. Contain myosin
    4. Are transversely disposed

36. LYMPH NODES:
1. Are surrounded by a capsule
2. Filter the blood
3. Are composed of nodules
4. Receive incoming lymph only at the hilus

37. IN THE ADULT HUMAN, HYALINE CARTILAGE IS FOUND IN THE:
1. Respiratory passages
2. Epiglottis
3. Metacarpo-phalangeal articulations
4. External ear

QUESTIONS 38–61:

THE GROUP OF QUESTIONS BELOW CONSISTS OF NUMBERED HEADINGS, FOLLOWED BY A LIST OF LETTERED WORDS OR PHRASES. FOR EACH HEADING SELECT THE *ONE* WORD OR PHRASE WHICH IS MOST CLOSELY RELATED TO IT.
NOTE: EACH CHOICE MAY BE USED *ONLY ONCE:*

38. Mitochondria
39. Attached ribosomes
40. Lysosomes
41. Golgi apparatus
42. Free ribosomes

A. Synthesis of protein for use by the cell
B. Synthesis of protein for export
C. Enzymes for cell respiration
D. Enzymes for cell destruction
E. Condensation of secretory product

43. Ciliated epithelium
44. Striated columnar epithelium
45. Simple squamous epithelium
46. Stratified squamous epithelium
47. Transitional epithelium

A. Gaseous exchange
B. Absorption
C. Protection
D. Distension
E. Removal of particulate matter

48. Epiphyseal plate
49. Epiphysis
50. Metaphysis
51. Diaphysis

A. Composed of cartilage
B. Formed from secondary ossification centre
C. Most active site of bone formation
D. Contains yellow marrow in the adult

52. Elastic cartilage
53. Hyaline cartilage
54. Fibrocartilage
55. Cancellous bone
56. Compact bone

A. Epiphysis of femur
B. Epiglottis
C. Pubic symphysis
D. Shaft of radius
E. Tracheal rings

57. Brachiocephalic artery
58. Medium size arteries
59. Arterioles
60. Capillaries
61. Veins

A. Prodominance of muscle in walls
B. Predominance of elastic in walls
C. No muscle in walls
D. Walls thick relative to size of lumen
E. Walls thin relative to size of lumen

QUESTIONS 62–101:

IN REPLY TO THE FOLLOWING QUESTIONS INDICATE WHETHER YOU THINK EACH STATEMENT IS *TRUE* OR *FALSE:*

TENDON:
62. Is a type of dense connective tissue adapted to weight-bearing
63. Is primarily a cellular tissue with a protective function
64. Has dense bundles of collagen all running in the same direction and causing the cells to be distributed in a linear manner
65. Has a rich blood supply when fully formed
66. Cannot undergo repair if severed

CARTILAGE:
67. Is an avascular tissue
68. Never forms a permanent structure
69. Increases in amount by both appositional and interstitial growth
70. Cannot survive if invaded by blood capillaries
71. Is everywhere covered by perichondrium

MACROPHAGES:
72. Produce antibody in response to antigenic stimulation
73. Develop from embryonic endodermal cells
74. Are actively phagocytic
75. Are conspicuous for the absence of lysosomes from their cytoplasm
76. Are added to throughout life by monocytes from the blood

ADIPOSE TISSUE:
77. Is derived from embryonic ectoderm
78. Has a rich blood supply
79. Receives a sympathetic nerve supply
80. Is a good insulator
81. Develops from embryonic mesenchyme

THE FIBROBLAST:
 82. Is a relatively inactive, mature connective tissue cell
 83. Has an abundance of cytoplasmic granules which obscure its nucleus when viewed with the light microscope
 84. Has abundant rough-surfaced endoplasmic reticulum
 85. Shows the features of an actively secreting cell when viewed with the electron microscope
 86. Secretes the fully-formed collagen fibre into the surrounding intercellular ground substance

MAST CELLS:
 87. Produce antibodies
 88. Commonly lie along the course of capillaries
 89. Contain large cytoplasmic granules
 90. Are responsible for the production of heparin
 91. Are the chief source of histamine

PLASMA CELLS:
 92. Are numerous in blood plasma
 93. Participate in the immune response
 94. Are found in abundance beneath wet epithelial surfaces
 95. Are absent from connective tissue underlying respiratory epithelium
 96. Are present in lymph nodes

SMOOTH MUSCLE CELLS:
 97. Have a single nucleus located in the widest part of the cell
 98. Contain myofilaments which are visible only with the electron microscope
 99. Remain constant in size throughout life
 100. Can be replaced by transformation of mesenchymal cells to muscle cells even in adult life
 101. Form tight junctions with each other to aid in the transmission of waves of contraction

INTRAMEMBRANOUS OSSIFICATION

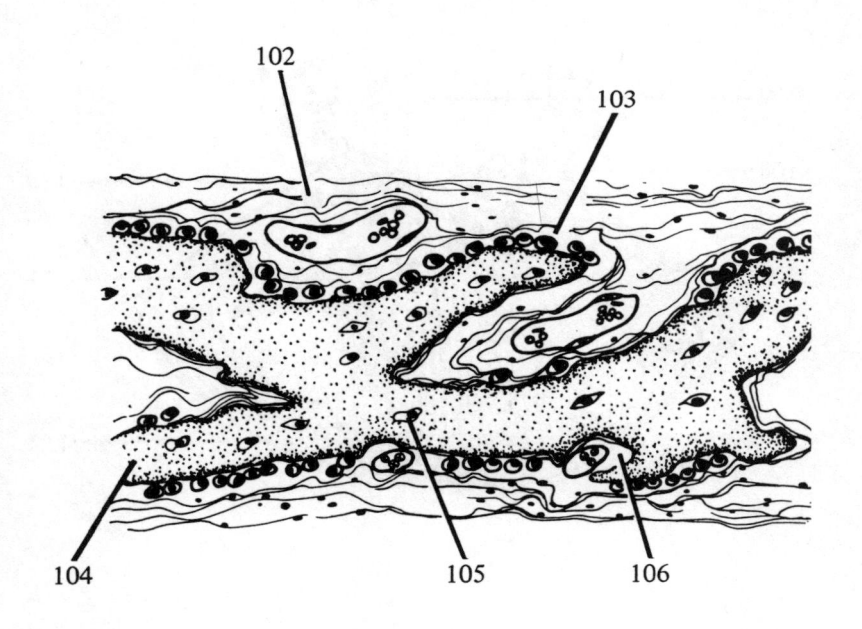

IDENTIFY THE NUMBERED STRUCTURES:

102
103
104
105
106

- A. Periosteum
- B. Osteocyte
- C. Howship's lacuna
- D. Osteoblast
- E. Chondroblast
- F. Bone
- G. Cartilage
- H. Capillary

ACINAR CELL OF PANCREAS

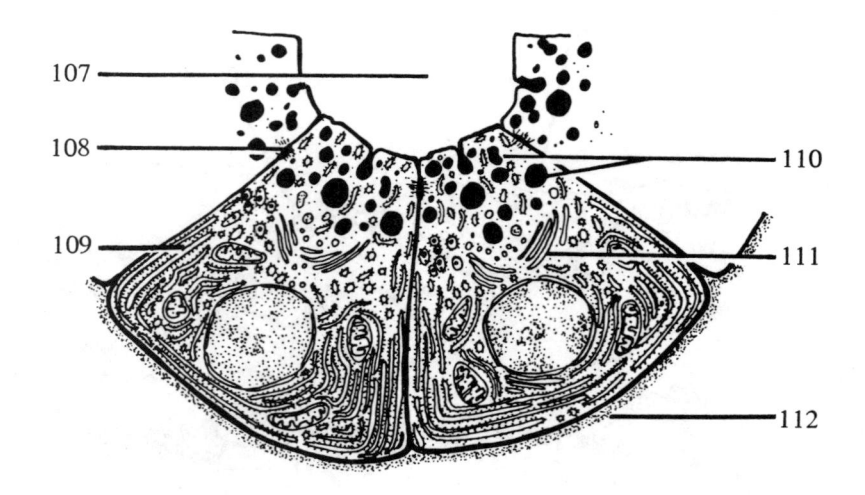

IDENTIFY THE NUMBERED STRUCTURES:

A.	Nuclear pore	F.	Acinar lumen
B.	Nucleolus	G.	Zymogen granules
C.	Mitochondrion	H.	Golgi complex
D.	Terminal bar	I.	Endoplasmic reticulum
E.	Microvillus	J.	Basement membrane

DEVELOPING BONE

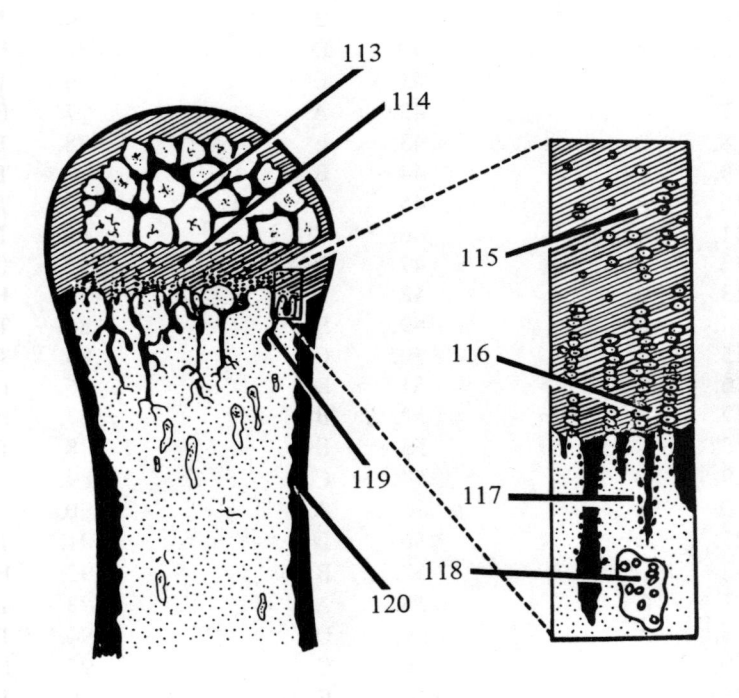

IDENTIFY THE NUMBERED STRUCTURES:

- A. Epiphyseal plate
- B. Calcified cartilage
- C. Perichondrium
- D. Bone marrow
- E. Subperiosteal bone
- F. Osteocyte
- G. Resting cartilage
- H. Trabecula
- I. Epiphyseal bone
- J. Osteoblast

ANSWERS

1.	D	36.	B	71.	F
2.	A	37.	B	72.	F
3.	B	38.	C	73.	F
4.	A	39.	B	74.	T
5.	A	40.	D	75.	F
6.	B	41.	E	76.	T
7.	C	42.	A	77.	F
8.	B	43.	E	78.	T
9.	C	44.	B	79.	T
10.	A	45.	A	80.	T
11.	C	46.	C	81.	T
12.	C	47.	D	82.	F
13.	B	48.	A	83.	F
14.	D	49.	B	84.	T
15.	B	50.	C	85.	T
16.	C	51.	D	86.	F
17.	A	52.	B	87.	F
18.	B	53.	E	88.	T
19.	D	54.	C	89.	T
20.	B	55.	A	90.	T
21.	B	56.	D	91.	T
22.	C	57.	B	92.	F
23.	C	58.	A	93.	T
24.	A	59.	D	94.	T
25.	B	60.	C	95.	F
26.	B	61.	E	96.	T
27.	A	62.	F	97.	T
28.	A	63.	F	98.	T
29.	D	64.	T	99.	F
30.	B	65.	F	100.	T
31.	A	66.	F	101.	T
32.	B	67.	T	102.	A
33.	A	68.	F	103.	D
34.	B	69.	T	104.	F
35.	B	70.	T	105.	B

106.	C	111.	H	116.	B
107.	F	112.	J	117.	J
108.	D	113.	I	118.	D
109.	I	114.	A	119.	H
110.	G	115.	G	120.	E

SECTION IX — GENERAL EMBRYOLOGY

QUESTIONS 1–17:

FOR EACH OF THE FOLLOWING MULTIPLE CHOICE QUESTIONS SELECT THE *ONE* MOST APPROPRIATE ANSWER:

1. THE CHROMOSOMAL FORMULA OF THE NORMAL HUMAN OVUM IS:
    A. 44 autosomes and 2 X chromosomes
    B. 45 autosomes and 1 X chromosome
    C. 22 autosomes and 1 Y chromosome
    D. 22 autosomes and 1 X chromosome
    E. 44 autosomes and 2 Y chromosomes

2. FOLLOWING OVULATION THE OVUM IS VIABLE FOR UP TO:
    A. One hour
    B. Twenty-four hours
    C. Forty-eight hours
    D. Four days
    E. Seven days

3. THE OVUM IS SURROUNDED BY A NON-CELLULAR, SECRETED LAYER KNOWN AS THE:
    A. Corona radiata
    B. Theca folliculi
    C. Zona pellucida
    D. Cumulus oophorus
    E. Stratum granulosum

4. THE SPERMATOZOON IS VIABLE WITHIN THE FEMALE GENITAL TRACT FOR UP TO:
   A. One hour
   B. Twenty-four hours
   C. Forty-eight hours
   D. Four days
   E. Seven days

5. THE USUAL NUMBER OF DAYS BETWEEN FERTILIZATION AND THE COMMENCEMENT OF IMPLANTATION IS:
   A. One
   B. Three
   C. Six
   D. Ten
   E. Fourteen

6. THE IMPLANTED EMBRYO WOULD BE EXPECTED ON GENETIC GROUNDS TO ELICIT A HOMOGRAFT RESPONSE FROM THE MOTHER. ITS IMMUNOLOGICALLY PRIVILEGED POSITION HAS BEEN ACCOUNTED FOR AS FOLLOWS:
   A. Embryonic tissues are not capable of producing antigens
   B. The trophoblast is only very feebly antigenic
   C. Maternal leucocytes enter the embryo and may induce acquired immunological tolerance
   D. The amniotic fluid confers immunological protection on the embryo
   E. The placental barrier blocks the transfer of antibodies from mother to fetus

7. THE UTERINE CAVITY IS OBLITERATED DURING PREGNANCY BY FUSION OF:
   A. Chorion frondosum and decidua basalis
   B. Decidua capsularis and decidua parietalis
   C. Amnion and decidua capsularis
   D. Amnion and chorion
   E. Endometrium and myometrium

8. THE "PLACENTAL BARRIER" BETWEEN MOTHER AND FETUS IS COMPOSED OF:
   A. Fetal capillary endothelium
   B. The trophoblast
   C. Maternal capillary endothelium
   D. All of the above
   E. A and B only

9. THE EMBRYO IS CALLED A FETUS AFTER THE:
   A. Third week
   B. Eighth week
   C. Twelfth week
   D. Sixteenth week
   E. Twenty-eighth week

10. AT THE ROSTRAL END OF THE PRIMITIVE STREAK LIES THE PRIMITIVE:
    A. Spot
    B. Node
    C. Notochord
    D. Plate
    E. Groove

11. AT THE CAUDAL END OF THE PRIMITIVE STREAK, ECTODERM AND ENDODERM MEET AT THE:
    A. Notochordal canal
    B. Coelom
    C. Cloacal membrane
    D. Neural groove
    E. Notochord

12. THE NEURAL CREST SEPARATES OFF FROM THE:
    A. Neural groove
    B. Neural fold
    C. Neural tube
    D. Neurenteric canal
    E. Neuropore

13. SEGMENTATION IS OBSERVED IN THE --------------- MESODERM:
    A. Paraxial
    B. Intermediate
    C. Lateral plate
    D. Splanchnic
    E. Branchial

14. THE COELOM IS FORMED IN THE EMBRYO BETWEEN THE LAYERS OF THE:
    A. Ectoderm
    B. Mesoderm
    C. Endoderm
    D. Somatopleure
    E. Splanchnopleure

15. THE MESENCHYMAL VERTEBRAE ARE FORMED FROM:
    A. Sclerotomes
    B. Myotomes
    C. Dermatomes
    D. Notochord
    E. Cartilage rudiments

16. THE FETAL MEMBRANES COMPRISE:
    A. Amnion
    B. Chorion
    C. Yolk sac
    D. All of the above
    E. A and B only

17. AT BIRTH, THE NORMAL UMBILICAL CORD CONTAINS:
    A. Two umbilical veins
    B. The vitelline duct
    C. The ductus venosus
    D. Two umbilical arteries
    E. The allantois

EARLY PLACENTA

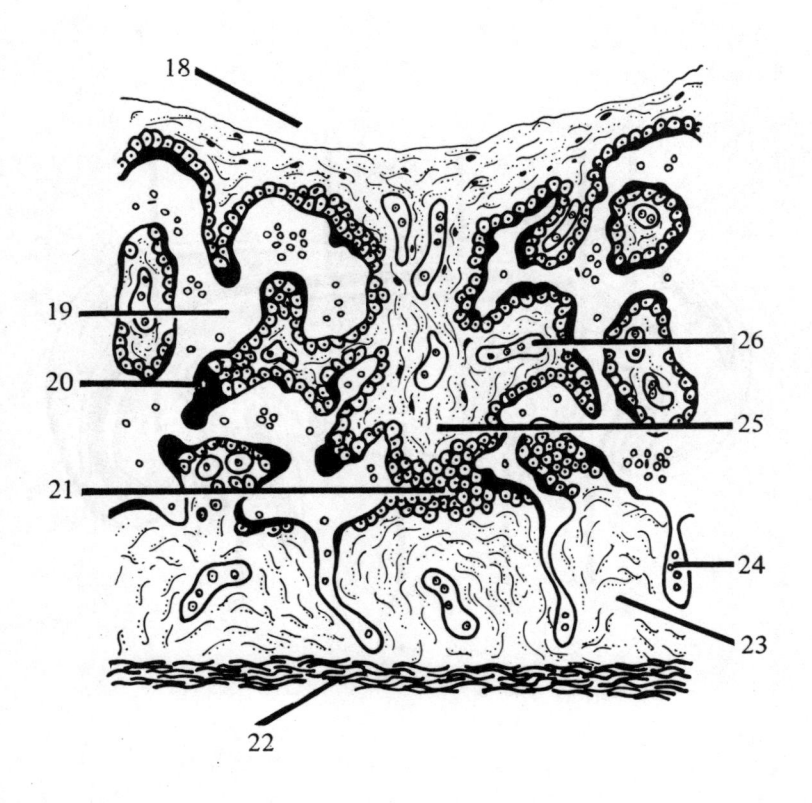

IDENTIFY THE NUMBERED STRUCTURES:

A.	Decidua basalis	I.	Extraembryonic coelom
B.	Intervillous space	J.	Intermediate mesoderm
C.	Embryonic mesoderm	K.	Cytotrophoblast
D.	Myometrium	L.	Primary villus
E.	Embryonic coelom	M.	Lymphatic vessel
F.	Syncytiotrophoblast	N.	Maternal blood vessel
G.	Vitelline vein	O.	Secondary villus
H.	Fetal blood vessel	P.	Extraembryonic mesoderm

189

25 DAY EMBRYO

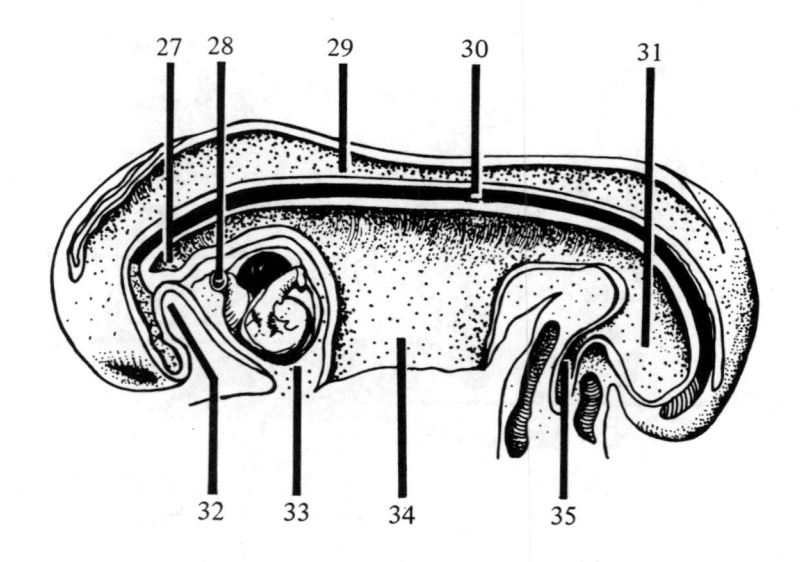

IDENTIFY THE NUMBERED STRUCTURES:

A. Foregut
B. Cloaca
C. Notochord
D. Rectum
E. Genital tubercle
F. Embryonic coelom
G. Vitelline duct
H. Neurenteric canal
I. Truncus arteriosus
J. Urogenital sinus
K. Neural tube
L. Stomodeum
M. Amnion
N. Neuropore
O. Allantois
P. Septum transversum

190

EMBRYO IN UTERO

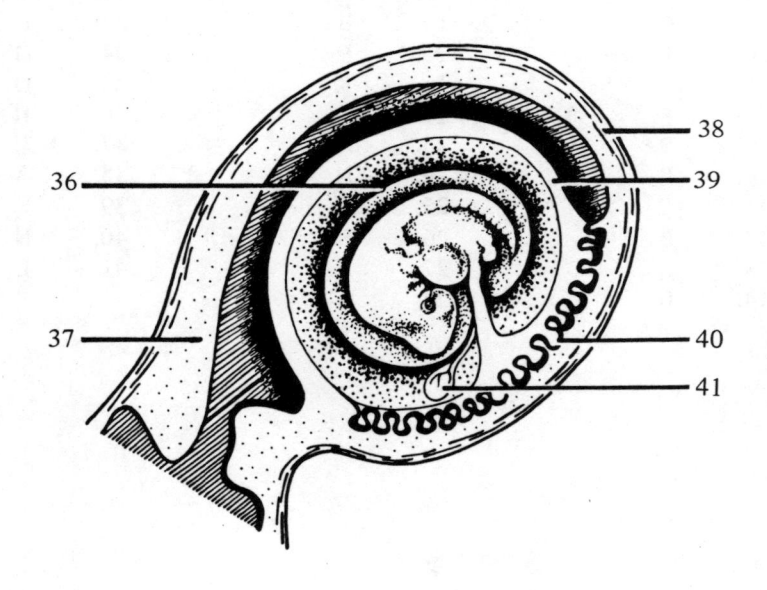

IDENTIFY THE NUMBERED STRUCTURES:

A.	Myometrium	H.	Amnion
B.	Embryonic endoderm	I.	Uterine cavity
C.	Ectoderm	J.	Decidua parietalis
D.	Intervillous space	K.	Extraembryonic coelom
E.	Decidua capsularis	L.	Yolk sac
F.	Umbilical cord	M.	Allantois
G.	Embryonic mesoderm	N.	Chorion frondosum

ANSWERS

1.	A	15.	A	29.	K
2.	B	16.	D	30.	C
3.	C	17.	D	31.	B
4.	C	18.	I	32.	L
5.	C	19.	B	33.	P
6.	B	20.	F	34.	G
7.	B	21.	K	35.	O
8.	E	22.	D	36.	H
9.	B	23.	A	37.	J
10.	B	24.	N	38.	A
11.	C	25.	P	39.	E
12.	B	26.	H	40.	N
13.	A	27.	A	41.	L
14.	B	28.	I		